T0336039

Carbon Nanotube and Graphene Device Physics

Explaining the properties and performance of practical nanotube devices and related applications, this is the first introductory textbook on the subject. All the fundamental concepts are introduced, so that readers without an advanced scientific background can follow all the major ideas and results. Additional topics covered include nanotube transistors and interconnects, and the basic physics of graphene. Problem sets at the end of every chapter allow readers to test their knowledge of the material covered and gain a greater understanding of the analytical skillsets developed in the text. This is an ideal textbook for senior undergraduate and graduate students taking courses in semiconductor device physics and nanoelectronics. It is also a perfect self-study guide for professional device engineers and researchers.

H.-S. Philip Wong is a Professor of Electrical Engineering at Stanford University, where he has worked since 2004. Prior to joining Stanford University, he spent 16 years at IBM T. J. Watson Research Center, Yorktown Heights, New York, where he held various positions from Research Staff Member to Senior Manager. He is a Fellow of the IEEE and his current research covers a broad range of topics, including carbon nanotubes, semiconductor nanowires, self-assembly, exploratory logic devices, nano-electromechanical devices, and novel memory devices.

Deji Akinwande is an Assistant Professor at the University of Texas, Austin, which he joined after receiving his Ph.D. from Stanford University in 2009. Prior to beginning his Ph.D., he was a circuit designer at Agilent Technologies, XtremeSpectrum/Freescale, and Motorola. He has published widely on carbon nanomaterials.

"An excellent and timely volume on the physics and applications of carbon nanotubes. A must read for students and researchers in this hot field."

Yuan Taur, UCSD

"This is the textbook that I have been aspiring to see for a long time. With excellent timing, the authors provide one that covers device physics of carbon nanotubes in a coherent, systematic way. The content is perfectly designed and formulated such that both students with little knowledge and researchers with hands-on experience in the field would find it extremely valuable. I would highly recommend this book to anyone who is interested in 'post-silicon' electronics."

Bin Yu, State University of New York, Albany

"I strongly recommend this text book to students, engineers and researchers who wish to build up their knowledge on carbon nanotube fundamentals and applications. They will extend their learning from materials technology and solid state physics to their applications in the fields of nanoelectronics and micro- nanosystems. The readers, interested by graphene and carbon nanotubes based devices, have the possibility to train themselves on the hottest topics and challenges which will pave the future of nanotechnology."

Simon Deleonibus, IEEE Fellow, CEA-LETI Chief Scientist and Research Director, MINATEC, Grenoble, France.

"This book is an excellent overview of carbon-based electronics, and in particular it provides the reader with an up-to-date and crisp description of the physical and electrical phenomena of carbon nanotubes, as well as a perspective on new applications enabled by this nanotechnology.

Both experts and students will enjoy reading this book, as it brings up to focus the important details of carbon solid-state physics to understand the ground rules of carbon transistors and the related nanoelectronic circuits. Moreover, from a global point of view, carbon electronics is a key nanotechnology supporting the continuous development of the information age in computing, sensing and networking."

Giovanni De Micheli, EPFL

"Excellent book covering all aspects of carbon nanotube devices from basic quantum physics in solids over material and device physics to applications including interconnects, field effect transistors and sensors. First complete book in an exciting new nanoelectronics field with great potential, intended for undergraduate and graduate students, researchers in the field and professional engineers, enabling them to get an insight in the field or to broaden their competence."

Cor Claeys, IMEC, Leuven, Belgium

Carbon Nanotube and Graphene Device Physics

H.-S. PHILIP WONG
Stanford University

DEJI AKINWANDE
University of Texas, Austin

CAMBRIDGE
UNIVERSITY PRESS

University Printing House, Cambridge CB2 8BS, United Kingdom

Published in the United States of America by Cambridge University Press, New York

Cambridge University Press is part of the University of Cambridge.

It furthers the University's mission by disseminating knowledge in the pursuit of education, learning and research at the highest international levels of excellence.

www.cambridge.org
Information on this title: www.cambridge.org/9780521519052

© Cambridge University Press 2011

This publication is in copyright. Subject to statutory exception and to the provisions of relevant collective licensing agreements, no reproduction of any part may take place without the written permission of Cambridge University Press.

First published 2011

A catalogue record for this publication is available from the British Library

Library of Congress Cataloguing in Publication data
Wong, Hon-Sum Philip, 1959–
Carbon nanotube and graphene device physics / H.-S. Philip Wong, Deji Akinwande.
   p.   cm.
Includes bibliographical references and index.
ISBN 978-0-521-51905-2
1. Nanotubes.   2. Graphene.   3. Optoelectronics–Materials.
4. Semiconductors–Materials.   I. Akinwande, Deji.   II. Title.
QC611.W86 2010
520'.5–dc22   2010039191

ISBN 978-0-521-51905-2 Hardback

Additional resources for this publication at www.cambridge.org/9780521519052

Cambridge University Press has no responsibility for the persistence or accuracy of URLs for external or third-party internet websites referred to in this publication, and does not guarantee that any content on such websites is, or will remain, accurate or appropriate.

Contents

Preface

Carbon nanotubes have come a long way since their modern rediscovery in 1991. This time period has afforded a great many scholars across the globe to conduct a vast amount of research investigating their fundamental properties and ensuing applications. Finally, after two decades, the knowledge and understanding obtained, once only accessible to select scholars, is now sufficiently widespread and accepted that the time is ripe for a textbook on this matter. This textbook develops the basic solid-state and device physics of carbon nanotubes and to a lesser extent graphene. The lesser coverage of graphene is simply due to its relative infancy, with a good deal of the device physics still in its formative stage.

The technical discourse starts with the solid-state physics of graphene, subsequently warping into the solid-state physics of nanotubes, which serves as the foundation of the device physics of metallic and semiconducting nanotubes. An elementary and limited introduction to the device physics of graphene nanoribbons and graphene are also developed. This textbook is suitable for senior undergraduates and graduate students with prior exposure to semiconductor devices. Students with a background in solid-state physics will find this book dovetails with their physics background and extends their knowledge into a new material that can potentially have an enormous impact in society. Scholars in the fields of materials, devices, and circuits and researchers exploring ideas and applications of nanoscience and nanotechnology will also find the book appealing as a reference or to learn something new about an old soul (carbon). Research into potential applications of carbon nanotubes has been progressing at a rapid pace. We have refrained from cataloging the continual progress in applications because it is simply impossible to keep up with the developments. Instead, we focus on the fundamental physics and principles so the reader can easily utilize them for the application at hand. Scattered throughout the book are many references that offer further coverage on the technical matters for the interested reader. However, owing to the restricted space there are many outstanding references that are not cited in this book. The avid reader will find much educational pleasure in perusing the technical literature from time to time, as this is a rapidly advancing field of research.

Our "carbon journey" started when one of us was at the IBM T.J. Watson Research Center in Yorktown Heights, New York. Richard Martel (now at Université de Montréal) and Phaedon Avouris were generous with sharing their knowledge; and the management – John Warlaumont and Tom Theis – fostered a nurturing intellectual environment within the confines of an industrial research laboratory. The kernel of this book started from sections of a graduate-level course at Stanford University, first taught in the Fall of 2005. We thank the students who asked the pertinent questions and offered suggestions for improvements. We also want to take this opportunity to thank all our reviewers that devoted their valuable time in improving the final product. We incorporated most of your feedback and apologize that the limited time prevented some feedback from coming through. Thanks to Tayo Akinwande (MIT), Phaedon Avouris (IBM), Zhihong Chen (IBM, Purdue), Azad Naeemi (Georgia Tech), Eric Pop (UIUC), Emmanuel Tutuc (University of Texas), Yin Yuen (Stanford), and Jerry Zhang (Stanford). Your efforts are sincerely appreciated. Now that the textbook is published, we continue to welcome any and all feedback. Errors and omissions of all sorts will no doubt be brought to light despite our best intentions to produce an error-free textbook. As such, we anticipate errata will be made available at Cambridge University Press website and updated as the case warrants. An online resource for instructors is also available at the website: www.cambridge.org/wong.

It has been a great pleasure writing this textbook on the device physics of carbon materials. For us, this book was a labor of love, and the adventure involved in developing the content along a unifying theme was a great enriching experience and sufficient reward in and of itself. We hope that all readers will similarly find great enrichment and understanding as they explore the pages of this book. Finally, we would like to thank our families – the Akinwande clan, Cecilia Mui, Amelia Wong, and Emily Wong – for their support and understanding.

Cheers

Deji Akinwande

Austin, Texas

H.-S. Philip Wong

Stanford, California

September 2010

1   Overview of carbon nanotubes

Nature is the origin of all things.

1.1   Introduction

Carbon is an old but new material. It has been used for centuries going back to antiquity, but yet many new crystalline forms of carbon have only recently been experimentally discovered in the last few decades. These newer crystalline forms include buckyballs, carbon nanotubes (CNTs), and graphene, where the latter two are illustrated in Figure 1.1. Furthermore, carbon nanotubes come in two major flavors, the single-wall and multi-wall varieties, as shown in Figure 1.1a and b respectively. The newer forms of carbon have significantly contrasting properties compared with the older forms of carbon, which are graphite and diamond.[1] In particular, they share in common a hexagonal lattice or arrangement of carbon atoms. In addition, CNTs and graphene occupy a reduced amount of space compared with their older siblings; hence, they are often referred to as reduced-dimensional or low-dimensional solids or nanomaterials for short. To give a comparative (order of magnitude) idea of the critical size scales of these nanomaterials, nanotubes are about 10 000 times thinner than human hair, and graphene is about 300 000 times thinner than a sheet of paper. The typical diameter of nanotubes range from about 1 to 100 nm, and graphene ideally has the thickness of a single atomic layer ($\sim$3.4 Å). Fundamentally, it is the combination of the reduced dimensions and the different lattice structure that leads to the fascinating properties unique to nanotubes and graphene.

The fascinating electronic properties are the main subject matter of this textbook. While both nanotubes and graphene will be formally studied, the primary focus will be on CNTs. This is because nanotubes have been actively studied since 1991, affording a relatively greater wealth of understanding compared with graphene, which have only recently enjoyed intense scrutiny (since 2005). The excitement and accumulated knowledge regarding nanotubes can be seen in the number of related articles published in the international journals of three major disciplines

---

[1] Quite fascinatingly, carbon produces the most esthetically beautiful of materials, diamond, and also arguably the ugliest of materials, graphite (the stuff found in pencils).

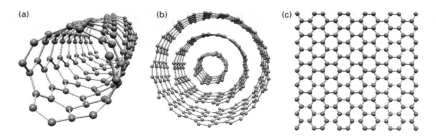

Fig. 1.1   Ball-and-stick models of CNTs and graphene: (a) single-wall nanotube, (b) multi-wall nanotube with three shells, and (c) graphene, which is a single sheet of graphite. The balls (spheres) represent the carbon atoms and the sticks (lines) stand for the bonds between carbon atoms.

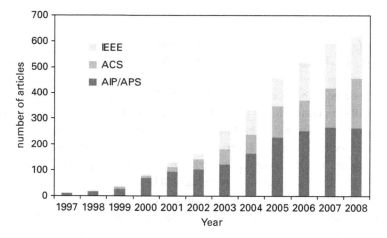

Fig. 1.2   Increase in the number of articles in the major international journals with the phrase "carbon nanotube" in the title. AIP/APS stands for the journals of the American Institute of Physics and American Physical Society respectively. ACS is the acronym for the journals of the American Chemical Society, and IEEE represents the journals of the Institute of Electrical and Electronic Engineers.

(physics, chemistry, and electrical engineering), which is reflected in Figure 1.2. Indeed, CNTs have enjoyed a somewhat exponential interest over the past decade, and the current gradual saturation in publications is a sign of the maturity of understanding about their properties. Accordingly, many diverse and novel applications have been explored, as can be seen in Figure 1.3, which is an indicator of the growing applications of CNTs.

This chapter will present a broad historical perspective on CNTs and a brief discussion of the common synthesis and characterization methods. It concludes with a note about non-CNT so that the reader is not left with the impression than only carbon atoms can form nanotubes. The discussion in this chapter is simply an

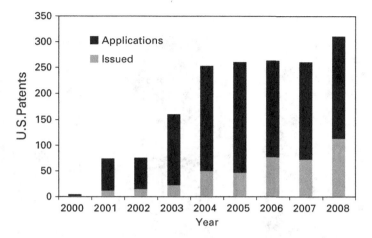

Fig. 1.3 United States patents issued and patent applications containing the phrase "carbon nanotube" in the patent abstract. In a sense, this is an indicator of the growing interest in applications of CNTs.

overview and is not required for any of the subsequent chapters. For an advanced technical overview from a different perspective, readers are encouraged to read some of the recent comprehensive reviews on CNTs, such as the one by Avouris *et al.*[2]

1.2 An abbreviated zigzag history of CNTs

The developments leading to the *discovery* of CNTs, or perhaps more fittingly their *accidental synthesis*, followed a somewhat jagged path of experimental research. Before proceeding any further, the reader should keep in mind that natural deposits of CNTs may well exist in some as yet undiscovered location(s). However, these locations (if they exist) have not been found, or at least no one has been actively searching for them. Moreover, the production of nanotubes may have similarly existed through some inadvertent process for a long time. A case in point is an interesting article by Indian scientists reporting on the synthesis of CNTs from oxidation of specially prepared carbon soot popularly known as *kajal* in South Asia, a substance that has been used as an eye-makeup as far back as the eighth century BC.[3] Within our scientific context, the historical discussion here will be limited to synthetic or man-made nanotubes that have clearly been identified by undeniable

[2] Ph. Avouris, Z. Chen and V. Perebeinos, Carbon-based electronics. *Nat. Nanotechnol.*, **2** (2007) 605–15.

[3] P. Dubey, D. Muthukumaran, S. Dash, R. Mukhopadhyay and S. Sarkar, Synthesis and characterization of water-soluble carbon nanotubes from mustard soot. *Pramana*, **65** (2005) 687–97.

Carbon nanotube   Carbon nanofiber   Carbon fiber

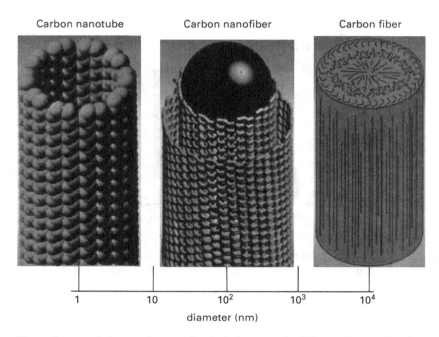

1	10	10^2	10^3	10^4

diameter (nm)

Fig. 1.4   Illustrative morphology and range of typical diameters for different filamental carbon.

evidence. The origins of these CNTs are ultimately rooted and intertwined with developments of their siblings, carbon nanofibers and fibers which are hair-like filaments. All three slender structures, namely nanotubes, nanofibers and fibers, can be categorized as filamental carbon for short, with dimensions and morphologies illustrated in Figure 1.4.

Historically, the intentional synthesis of carbon fibers can be traced back to the time of Thomas Edison and the beginnings of electrical lighting (late 1800s), when this material enjoyed a growing application as filaments in the commercially successful incandescent light bulb. Indeed, a US patent was issued in 1889 to two British scientists documenting the detailed synthesis of carbon filaments or fibers by what is now known as the chemical vapor deposition (CVD) method. The first page of the patent is shown in Figure 1.5.

Remarkably, the patented CVD recipe, which employs iron to catalyze the thermal decomposition of a mixture of hydrogen and hydrocarbon gases (methane and ethylene are mentioned in the patent as preferable) leading to the growth of carbon fibers, remains the most popular method of synthesizing nanofibers and nanotubes even after a century's worth of research in carbon materials. Some of the highlights in the patent schematic (all labels refer to Figure 1.5) include a gas outlet (labeled k) to bake out and evacuate any moisture/steam in the chamber before growth and an inlet (c) to allow the flow of the gases through a small opening (b^x) for subsequent carbon synthesis at high temperatures. The carbon fibers grow from

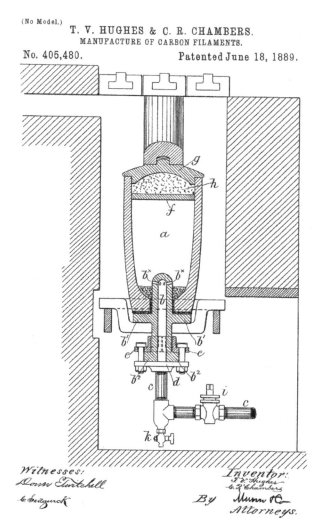

(No Model.)

T. V. HUGHES & C. R. CHAMBERS.
MANUFACTURE OF CARBON FILAMENTS.

No. 405,480.   Patented June 18, 1889.

Witnesses:

Inventor:

By   Attorneys.

Fig. 1.5   Patent drawing of Hughes and Chambers for the invention of a catalytic high-temperature technique for growing carbon fibers or filaments in a CVD chamber under atmospheric conditions.

the bottom and the walls of the iron crucible to fill the $\sim$5-inch tall chamber and were mentioned to be of low electrical resistance and high density. At the time, the role of iron as a catalyst was not recognized in the patent. Now we know that iron is a particularly efficient catalyst for filamentary carbon synthesis from methane. A sequence of clay-like layers (f/h) is used to trap the gaseous by-products which are removed after synthesis. Essentially the same procedure is employed today for the basic CVD synthesis of nanotubes, typically employing a quartz chamber and patterned iron nanoparticles on a substrate for localized directed growth of CNTs.

Fig. 1.6   TEM images of what appears to be multi-wall CNTs. Adapted from the 1952 article by
Radushkevich and Lukyanovich.[5] The upper and lower multi-wall CNTs have diameters
of 50 nm and 100 nm respectively.

With the development of the electron microscope in the 1930s, scientists had a
new, extremely useful (and expensive) tool to play with. They immediately began
to take a close look at the structure of nature at length scales that had been previously inaccessible.[4] In 1952, Russian scientists reported intriguing transmission
electron microscope (TEM) images of filamental carbon structures synthesized
from the iron-catalyzed decomposition of carbon monoxide at temperatures in the
range of 400–700 °C.[5] The TEM images (Figure 1.6) are believed to be the first
to show what appears to be (what we now call) multi-wall nanotubes among the
other filamental and amorphous carbon products described in the article. They
recognized the catalytic properties of iron in facilitating the growth of tubular
carbon, and also discussed the initial formation of iron carbide at the base of the
tubes followed by the subsequent growth of carbon filaments, a theory which is
widely accepted today. Perhaps partly due to *cold war* difficulties in rapidly disseminating new knowledge at that time, British materials researchers (apparently
unaware of the Russian article) independently announced similar filamental observations resulting from thermal decomposition of carbon monoxide in the presence
of iron a year later.[6] Their work appears to have been motivated by the need to
understand the degradation of ceramic bricks used in blast furnaces. Blast-furnaces
are furnaces used for the production of industrial metals such as iron, and bricks
are traditionally used in the furnace lining. The deposition of amorphous and filamental carbon on the bricks is entirely undesirable in a blast furnace because
it compromises the material integrity of the bricks, resulting in a serious reliability issue. This had been a major problem they were investigating. They also
speculated that carbon filaments likely accumulate in domestic chimneys, because

---

[4] Optical microscopes can provide a magnification up to about 1000 times, while electron
microscopes can offer a magnification of about a million times, corresponding to sub-nanometer
resolution.

[5] L. V. Radushkevich and V. M. Lukyanovich, O strukture ugleroda, obrazujucegosja pri
termiceskom razlozenii okisi ugleroda na zeleznom kontakte (On the structure of carbon formed by
the thermal decomposition of carbon monoxide to the contact with iron). *Zh. Fis. Khim. (Russ. J.
Phys. Chem.)*, **26** (1952) 88–95.

[6] W. R. Davis, R. J. Slawson and G. R. Rigby, An unusual form of carbon. *Nature*, **171** (1953) 756.

similar processes occur. Accordingly, it is certainly plausible that the unintentional production of carbon fibers and tubes may have been going on for a very long time, say for at least as long as chimneys have been around. This is another reason why we prefer to focus on the scientific investigations of carbon morphologies.

Inspired by the reports from the Russian and British researchers, Hofer *et al.* followed up on the suggestion from the former that similar carbon filaments might be produced using cobalt or nickel catalysts instead of iron. In 1955, they were successful in reproducing the synthesis of carbon filaments from carbon monoxide using either cobalt or nickel catalysts.[7]

So far, the electron microscope had been an indispensable tool for elucidating the morphology of the different carbon filaments and for bringing their general structural properties to light, leading to a renewal in scientific interest. However, no industrial application had been developed at that time, as tungsten had largely replaced carbon as filaments in the lighting industry by 1910.[8] Fortunately, all this changed by the 1960s, largely due to the pioneering work of Roger Bacon, an American scientist who was then with the National Carbon Company, a fitting name for a corporation devoted to applications of carbon materials. While studying the properties of graphite in an arc-discharge furnace under extreme conditions (close to its triple point, temperature ~3900 K, pressure ~92 atm), he discovered the formation of carbon nanofibers with somewhat different (but similar) morphology than had been previously observed.[9] The reported TEM images seem to show that the carbon nanofibers are composed of concentric cylindrical layers of carbon similar to a multi-wall nanotube but with a length-dependent diameter ranging from sub-micrometers to over 5 μm, leading Roger Bacon to propose a scroll model for the morphology for the fibers. Subsequent work by others has also shown that nanofibers can have a morphology similar to a stack of cones (or a stack of paper cups),[10] which is close to the structure of a paper scroll (see Figure 1.7).

Roger Bacon was able to optimize the growth conditions to yield high-performance polycrystalline carbon fibers with outstanding mechanical properties (Young's modulus ~700 GPa, tensile strength ~20 GPa) which were notably much superior to steel (Young's modulus ~200 GPa, tensile strength ~1–2 GPa), while retaining room-temperature resistivity (~65 μΩ cm) comparable to that of crystalline graphite.[9] In the following decade significant progress was made in commercializing carbon fiber technology. For example, around 1970 several high-profile review and news articles had been published that discussed the comparative

---

[7] L. J. E. Hofer, E. Sterling and J. T. McCartney, Structure of the carbon deposited from carbon monoxide on iron, cobalt and nickel. *J. Phys. Chem.*, **59** (1955) 1153–5.

[8] Remarkably, more than a century, later, tungsten continues to be the ubiquitous choice for filaments in incandescent light bulbs.

[9] R. Bacon, Growth, structure, and properties of graphite whiskers, *J. Appl. Phys.*, **31** (1960) 283–90.

[10] A case in point is: M. Endo *et al.*, Pyrolytic carbon nanotubes from vapor-grown carbon fibers, *Carbon*, **33** (1995) 873–81.

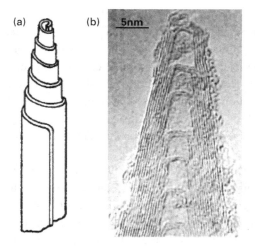

Fig. 1.7   Morphology of carbon nanofibers. (a) Scroll model proposed by Roger Bacon. Reprinted
with permission from Ref. 9. Copyright (1960), American Institute of Physics.
(b) High-resolution TEM image of a nanofiber showing a hollow cone-like structure with
multiple walls. Adapted from Endo *et al.*[10] Copyright (1995), with permission from
Elsevier.

benefits of carbon fibers and the new emerging applications.[11] If we consider
the late 1800s as the birth of carbon fibers driven by light-bulb applications,
then the 1960s can be seen as the *first renaissance* (a second one will be mentioned later) for fiber technology driven by new applications. The new applications
were grounded on taking advantage of the mechanical (strength, stiffness, light-
ness) and high-temperature properties of the fibers to manufacture composites (reinforced plastics, metals, glass, etc.) with tailor-made engineered performance for
a variety of industries, including vehicle parts and engines for the aerospace, space,
and automotive industries. Making filamental carbon was once again big business
with commercial ramifications. To tap their full commercial potential, much of
the subsequent development over the next few decades was in lower cost, higher
volume production of carbon fibers.

Along this line, Morinobu Endo, then a graduate student in Japan, started working on different recipes to synthesize lower cost carbon fibers in the early 1970s. At
that time, owing to the growing commercialization, a wide variety of manufacturing techniques had been developed. In France as a visiting scholar, he was exploring
the growth of carbon fibers on a substrate, based on thermal decomposition of a
gas mixture of benzene and hydrogen, when by chance he observed both single-
wall and multi-wall CNT (Figure 1.8). This was a total accidental rediscovery of

[11] See: New Materials make their mark. *Nature*, **219** (1968) 875; R. W. Chan and B. Harris, Newer
forms of carbon and their uses. *Nature*, **221** (1969) 132–41; and R. Jeffries, Prospects for carbon
fibres. *Nature*, **232** (1971) 304–7.

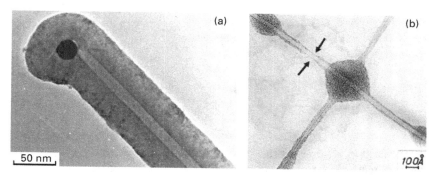

Fig. 1.8 Electron microscope images of multi-wall and single-wall hollow tubes recorded in 1976. (a) A thick nanotube with a hollow core corresponding to a multi-wall CNT with diameter of $\sim$70 nm. (b) The arrows are pointing to a section of a hollow tube with a single shell with diameter of $\sim$4 nm. Courtesy of Morinobu Endo. Reprinted from Ref. 13. Copyright (1976), with permission from Elsevier.

hollow carbon tubes brought about by a practical desire to hasten the cleaning of the substrate for later reuse, so as to increase the throughput of the CVD system. This is very much similar to baking cake in a pan in an oven. After baking, the pan has to be cleaned for subsequent baking runs. If the cleaning takes too long, one can imagine then that only so many baking runs are possible in a given time. Morinobu Endo was keen on reducing the time for cleaning the substrate (typically consumed about two and a half days) in order to increase the number of experiments he could run. By simply cleaning the blackened substrate with sandpaper, he afterwards noticed the growth (at 1100 °C) of what he termed hollow tubes (which we now call CNTs) facilitated by the catalytic iron particles unknowingly left behind from the sandpaper on the substrate.[12] Apparently unaware of the prior work of Raduskevich and Lukyanovich,[5] Endo and coworkers performed detailed studies elucidating the role of the iron as a catalyst, the crystalline structure of the hollow tubes and their long concentric shells, and even reported the first recorded image of a hollow tube with a single shell (see Figure 1.8b).[13] The single-wall CNT was a momentary curiosity and was all but forgotten shortly after because interest then was in carbon fibers which were increasingly well developed for use in the several industries employing composite materials. There was no study of the non-mechanical properties of the CNTs, such as their electrical, optical, and sensor properties, that might have shed light on new, interesting applications. That had to wait for the second renaissance in filamental carbon science about 15 years later.

[12] He was greatly inspired by his French supervisor (Professor Agnes Oberlin) to "see what is really there, not what you would like to see."

[13] A. Oberlin, M. Endo and T. Koyama, Filamentous growth of carbon through benzene decomposition. *J. Cryst. Growth*, **32** (1976) 335–49.

Fig. 1.9   Ball-and-stick model of C_{60} or buckyball. The balls (spheres) represent the carbon atoms and the sticks (lines) stand for the bonds between carbon atoms.

A major breakthrough occurred in 1985 with the *experimental discovery* of a zero-dimensional (relatively speaking, compared with larger forms of carbon) allotrope of carbon named C_{60} (or *buckyball* or *buckminsterfullerene* after Richard Buckminster "Bucky" Fuller, the American architect famous for geodesic domes) by the discoverers Kroto, Curl, Smalley, and coworkers at Rice University in Texas.[14] This discovery was completely by chance, as the chemists were performing experiments to simulate the conditions of carbon nucleation and formation in giant stars (much larger than the Sun) and interstellar space. In the course of these experiments they managed to observe C_{60}, which resembles a soccer ball. Indeed, in the landmark article, the authors showed an image of a soccer ball (notably pointing out it was on Texas grass) to illustrate the structure of C_{60}. A ball-and-stick model is shown in Figure 1.9. This discovery renewed focus on low-dimensional crystalline allotropes of carbon, and speculation about the possibility of single-wall CNTs had been swirling by the late 1980s (the previously reported images of single-wall and multi-wall hollow tubes had largely remained unnoticed).[15]

By 1990 there was a great deal of excitement, and at the Fall 1990 MRS (Materials Research Society) meeting in Boston, Kroto strongly urged Iijima (an electron microscope expert at NEC laboratories with a long history in carbon research) to work on buckyballs. Immediately, Iijima started conducting experiments aimed at elucidating the growth mechanism of buckyballs using an arc-discharge method common to fullerene synthesis. Within a year, by serendipity he noticed elongated hollow structures in the carbon soot which were multi-wall CNTs, including a double-wall nanotube where the walls can be clearly seen (Figure 1.10).[16]

[14] H. W. Kroto, J. R. Heath, S. C. O'Brien, R. F. Curl and R. E. Smalley, C_{60}: Buckminsterfullerene. *Nature*, **318** (1985) 162–3. Kroto, Curl, and Smalley subsequently won the Nobel Prize in Chemistry in 1996. The Nobel Prize is limited to a maximum of three recipients; as such, the then graduate student co-authors (Heath and O'Brien) did not participate in the award.

[15] M. S. Dresselhaus, G. Dresselhaus, and Ph. Avouris (editors), *Carbon Nanotubes: Synthesis, Structure, Properties, and Applications* (Springer-Verlag, 2001), Chapter 1.

[16] S. Iijima, Helical microtubules of graphitic carbon. *Nature*, **354** (1991) 56–8.

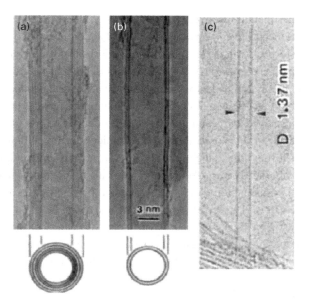

Fig. 1.10   High-resolution TEM images of multi-wall and single-wall CNTs observed by Sumio
Iijima in 1991 and 1993 respectively. (a) Multi-wall, (b) double-wall, and (c) single-wall
CNT. The first two images are adapted from Iijima,[16] and the latter image is from Iijima
and Ichihashi.[18] Reprinted by permission from Macmillan Publishers Ltd: copyright
(1991) and (1993).

This extends the run of accidental discoveries related to low-dimensional carbon
allotropes. The observation immediately had a high impact, in part due to the broad
dissemination that comes with a widely circulating journal. It heralded the entry
of many scientists into the field of crystalline carbon fibers, further encouraged
by initial theoretical studies of several condensed-matter groups reporting their
fascinating electronic properties, especially for the single-wall variety.[17] Indeed,
in 1992, the terms *armchair, zigzag*, and *chiral* had been introduced to characterize
the electro-physical properties of single-wall nanotubes. They were predicted to
be electrically metallic or semiconducting, depending on the diameter and chirality. The theoretical study of single-wall CNTs was distinct from earlier works,
which focused mostly on the mechanical and thermal properties of carbon fibers.
The properties of carbon fibers which have a polycrystalline structure are not as
beautiful and as interesting from a physics point of view as are the properties of a
well-ordered perfectly symmetric hollow nanotube.

[17] See: J. W. Mintmire, B. I. Dunlap and C. T. White, Are fullerene tubules metallic? *Phys. Rev. Lett.*,
68 (1992) 631–4; N. Hamada, S.-i. Sawada and A. Oshiyama, New one-dimensional conductors:
graphitic microtubules. *Phys. Rev. Lett.*, **68** (1992) 1579–81; and R. Saito, M. Fujita, G.
Dresselhaus and M. S. Dresselhaus, Electronic structure of graphene tubules based on C_{60}. *Phys.
Rev. B*, **46** (1992) 1804–11.

The actual experimental discovery, or shall we say *rediscovery*, of nanotubes with single walls occurred a year after the initial theoretical studies and was reported in back-to-back articles by two separate groups (by Iijima and Ichihashi at NEC and by Bethune *et al.* at IBM).[18] The rediscovery of single-wall CNTs coupled with the prediction and subsequent confirmation of their tunable electrical and optical properties is what has ushered in the ongoing second renaissance in filamental carbon science and technology. The intense interest, initially driven by considerations for electronic applications in the semiconductor technology industry,[19] has grown substantially to include many new promising applications, such as electronics on flexible substrates, all kinds of biological/chemical/physical sensors, supercapacitors, hydrogen storage materials, nanoprobes, and electron-emission sources to name a few. The legacy application of carbon fibers in engineered composites has also experienced a resurgence using nanotubes because single- or multi-wall nanotubes boast even higher mechanical strength than nanofibers. The practical value and economic impact of CNTs is potentially very large, prompting Iijima to remark jokingly at a 1997 speech to the Royal Institution in England that "one day, sir, you may tax it."[20]

In terms of fundamental science, the intense activity surrounding CNTs has stimulated another active branch of carbon research, namely that of *graphene*, which is an isolated two-dimensional single-atomic plane of graphite. While graphite has been known for centuries (large deposits were found in the 1500s in Cumbria, England), and the basic theoretical study of graphite and graphene reported as early as 1948, the experimental study of *free-standing graphene* devices actively commenced with the ground-breaking work of Novoselov, Geim, and coworkers in 2005.[21] Graphene will be formally introduced and studied in Chapter 3.

In summary, *carbon is a new but old material*. The long history of CNTs traces back to developments concerning carbon fibers which were initially synthesized for light-bulb applications in the late 1800s but largely replaced by tungsten by 1910. Carbon fibers later experienced an industrial renaissance in the early

---

[18] See: S. Iijima and H. Ichihashi, Single-shell carbon nanotubes of 1-nm diameter. *Nature*, **363** (1993) 603–5; and D. S. Bethune, C. H. Klang, M. S. de Vries, G. Gorman, R. Savoy, J. Vazquez and R. Beyers, Cobalt-catalysed growth of carbon nanotubes with single-atomic-layer walls. *Nature*, **363** (1993) 605–7.

[19] Notably, the first carbon nanotube field-effect transistors were demonstrated in 1998 (around the same time as nanotube research began to experience explosive growth in publications). See: S. Tans, A. R. M. Verschueren and C. Dekker, Room-temperature transistor based on a single carbon nanotube. *Nature*, **393** (1998) 49–52; and R. Martel, T. Schmidt, H. R. Shea, T. Hertel and Ph. Avouris, Single- and multi-wall carbon nanotube field-effect transistors, *Appl. Phys. Lett.*, **73** (1998) 2447–9.

[20] The original quote was by the celebrated British scientist Michael Faraday (an all around genius with no formal education), who gave such a response in 1850 when asked by the British minister of finance about the practical value of electricity.

[21] K. S. Novoselov, A. K. Geim, S. V. Morozav, D. Jiang, M. I. Katselson, I. V. Grigorieva, S. V. Duboros and H. A. Firsov, Two-dimensional gas of massless Dirac fermions in graphene. *Nature*, **438** (2005) 197–200.

1970s as composite materials as their mechanical properties became well understood. The first recorded images of multi-wall nanotubes were by Raduskevich and Lukyanovich in 1952, and that of a single-wall nanotube was by Endo and coworkers in 1976. However, the applications then were focused on carbon-based composite materials, as the fascinating electronic properties of nanotubes were not known, as such, consigning them to obscurity. It was not until the reappearance of CNTs due to the work of Iijima in 1991 that heralded the current rise in carbon science and technology. This ongoing renaissance in carbon nanomaterials is fueled by the diverse and tunable electronic and optical properties of CNTs for a broad range of novel applications.

1.3   Synthesis of CNTs

The current methods of synthesizing CNTs essentially require the pyrolysis or thermal decomposition of an appropriate carbon source, such as hydrocarbons or carbon monoxide. As such, these are high-temperature processes in a sealed chamber at a controlled pressure and temperature, typically employing metal catalysts to enhance the yield. In some methods, such as CVD, metal catalysts are necessary for the growth of carbon nanotubes. The favored metal catalysts are nickel, cobalt, iron, or a mixture of these elements. However, a recent study has determined that a much wider range of metals can serve as catalysts for nanotube growth, albeit with results that at best compare with those obtained from the favored metal catalysts.[22] Metal nanoparticles appear to serve as both nucleation sites for the vapor molecules and also as catalysts to reduce the activation energy needed for the decomposition of the hydrocarbon vapor. Nanotubes can also grow on rough metal films, where the ridges of the corrugated surface act like catalytic nanoparticles. The specific growth conditions can be optimized for either single-wall or multi-wall nanotubes. A comprehensive understanding of nanotube growth mechanisms is still being investigated.

All synthesis methods produce a variety of nanotubes with a distribution in diameter, invariably leading to a distribution in electronic and solid-state properties. Indeed, the synthesis of CNTs with a specific diameter or chirality (to be discussed in Chapter 4) is perhaps the *holy grail* of nanotube science. This is a matter of further research with wide-ranging implications for both fundamental science and the accelerated development of industrial applications. At present, it is possible to employ some form of post-synthetic treatment of the CNTs to sort the nanotubes, in order to achieve a narrower distribution with more uniform properties.[23]

[22] D. Yuan, L. Ding, H. Chu, Y. Feng, T. P. McNicholas and J. Liu, Horizontally aligned single-walled carbon nanotube on quartz from a large variety of metal catalysts. *Nano Lett.*, **8** (2008) 2576–9.

[23] M. C. Hersam, Progress towards monodisperse single-walled carbon nanotubes. *Nat. Nanotechnol.*, **3** (2008) 387–94.

What follows in this section is a survey of the three popular techniques for synthesizing both single-wall and multi-wall nanotubes. As such, the narrative is an elementary overview (i.e. lacking in detail) to provide the reader with a basic idea of the different methods for synthesizing CNTs. For a more comprehensive discussion of synthesis methods, the reader is encouraged to consult any one of the several wonderful texts available, including a recent book by Peter Harris.[24]

For the record (*circa* 2009), the smallest synthesized isolated single-wall CNT has a diameter of ~0.4 nm, in agreement with theoretical prediction.[25] The longest individual single wall grown horizontally on a substrate is ~40 mm.[26] Also, vertical forests of CNTs can routinely reach heights of several millimeters.

Chemical vapor deposition (CVD)

The CVD synthesis of CNTs occurs in a sealed high-temperature furnace (~500–1000 °C) where metal particles (frequently iron, nickel, cobalt, or their oxides which can be reduced before growth) catalyze the chemical vapor decomposition of a carbon-containing gas such as methane, ethylene, acetylene, carbon monoxide, etc. at atmospheric or sub-atmospheric pressures. Typically, the metal catalyst is deposited uniformly or selectively on a substrate, or they can also be introduced in the gas phase and in such cases are termed *floating catalysts*. Growth is initiated when the gases are introduced into the CVD chamber, which is typically made of high-temperature quartz. The carbon carrier gas nucleates on the surface of the metal particles, which catalyze the dissociation of the carbon gas with the freed carbon atoms arranging in a cylindrical manner to form a nanotube. After growth, the furnace is cooled to low temperatures (<300 °C) before exposing the CNTs to air in order to prevent any oxidative damage to the nanotubes. The catalyst particle is usually found as a carbide at the base or tip of the nanotube, depending on the adhesion of the particle to the substrate surface. In some cases, the particle can be somewhere around the middle of the nanotube, indicating that carbon atoms extend out in opposite directions of the catalytic particle. The diameter distribution of the nanotubes produced is often related to the statistical distribution of the particle sizes, and the length of the nanotube is dependent on the duration of growth. Growth rates can vary by many orders of magnitude, from several nanometers per minute to hundreds of micrometers per minute. Elucidation of the growth chemistry is a matter of ongoing research.

The CVD process is very flexible and can be optimized to produce randomly oriented nanotubes as well as horizontally and vertically aligned nanotubes. Dense

[24] P. J. F. Harris, *Carbon Nanotube Science: Synthesis, Properties, and Applications* (Cambridge University Press, 2009).

[25] N. Wang, Z. K. Tang, G. D. Li and J. S. Chen, Single-walled 4 Å carbon nanotube arrays, *Nature*, **408** (2000) 50–1.

[26] L. X. Zheng, M. J. O'Connell, S. K. Doorn *et al*. Ultralong single-wall carbon nanotubes. *Nat. Mater.*, **3** (2004) 673–6.

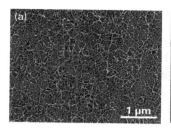
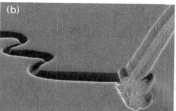
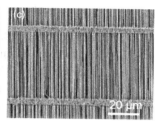

Fig. 1.11   Scanning electron microscope images of single-wall CNTs on oxidized silicon wafers. (a) A random network of nanotubes, (b) a CNT vertical forest $\sim$10 μm tall (courtesy of G. Zhang and H. Dai, Stanford University), and (c) horizontally aligned arrays of CNTs.

random networks of CNTs usually occurs in regions with a large density of metal catalysts uniformly dispersed on the substrate (Figure 1.11a). For very high-yield growth conditions (of the order of one CNT/particle), a dense dispersion of the catalysts can lead to vertically aligned CNT forests due to the crowding effect of the nanotubes. Alternatively, the CVD can be plasma enhanced (PECVD), resulting in a vertical electric field which, under optimum conditions, facilitates the growth of nanotube forests perpendicular to the substrate (Figure 1.11b). Additionally, the energy provided by the plasma allows the growth temperature to be much lower, especially for multi-wall nanotubes, and possibly down to room temperature for carbon nanofibers. For horizontally aligned straight nanotubes, the catalyst particles are dispersed on lithographically patterned islands or strips on the substrate. However, the orientation of the alignment is random, with no a priori knowledge of the direction of growth, resulting in very low yield for nanotube devices with pre-patterned structures at specific locations on the substrate. Fortunately, perfectly aligned nanotubes with deterministic orientation are achievable on the surface of certain crystalline substrates, such as quartz and sapphire (Figure 1.11c).

Arc discharge

The generation of an electric arc (a current of ionized gas) of carbon atoms can be achieved by applying a very high electric field between two graphite rods at high temperatures. The carbon plasma is attracted to the negative electrode (cathode), and under optimum conditions, typically employing the favorite metal catalysts (e.g. cobalt, nickel, iron), the vapor will condense to grow CNTs. The arc-discharge synthesis of nanotubes is typically accomplished in a water-cooled helium chamber at around atmospheric/sub-atmospheric pressures. The graphite electrodes are separated by a gap of the order of millimeters with a gap voltage in the vicinity of 20 V. Traditionally, the arc-discharge method was used to produce buckyballs; as such, there is often a variety of carbon structures that are simultaneously produced at the same time as nanotubes. Hence, the yield of nanotubes is normally <40% by weight. Solution processing is necessary in order to harvest and purify the CNTs.

Laser ablation

Ablation means the removal of matter from the surface of a substance through some process; for example, by the application of a laser. Hence, as the phrase implies, the synthesis of CNTs by laser ablation employs a laser to vaporize the surface of a graphite target in a reactor preheated to high temperatures. Normally, the target is not pure, but a graphite composite loaded with the common CNT catalytic metals (a cobalt–nickel mix is popular) to improve the yield. The effect of the laser is to significantly increase the target surface temperature beyond the furnace temperature and vaporize the target. Afterwards, the carbon vapor condenses to produce nanotubes on a water-cooled surface of the reactor. The laser ablation method typically produces mostly single-wall nanotubes with a relatively narrow diameter. The nanotubes are generally in a bundle held together by van der Waals inter-nanotube forces, and can be subsequently processed for greater purity. Most of the early CNT transistors were made with laser-ablated CNTs.

1.4   Characterization techniques

Invariably, the characterization of CNTs is a practical matter that any experimentalist must be well skilled in order to determine the CNT physical dimensions, structural quality, and electronic character (metal or semiconducting). The three routine techniques of utmost utility are atomic force microscopy (AFM), scanning electron microscopy (SEM), and Raman spectroscopy. All three techniques provide a relatively fast method to confirm the presence of CNTs and characterize them as is, before any further processing. AFM and SEM provide a visual image obtained by scanning the substrate surface containing CNTs, while Raman spectroscopy is an optical method that probes the lattice vibrations of materials at the surface from where a characteristic peak corresponding to nanotubes can be easily identified in the Raman spectra. In addition to these three techniques, a TEM is also very useful in characterizing the dimension and structure of CNTs, though somewhat complex sample preparation or processing is usually required. In general, a TEM is the most accurate method for measuring the diameter of a nanotube. For typical routine characterizations, the nanoscale spatial resolution of standard AFM and SEM means that an individual nanotube can be identified and its dimensions determined with reasonable accuracy. On the other hand, owing to the micrometer spot size of standard Raman spectrometers, one can only determine if nanotubes are indeed present. The quantity or density of the nanotubes cannot be ascertained with reasonable confidence. However, one of the major advantages of Raman spectroscopy is in probing the collective properties of an ensemble of nanotubes that produces a qualitative picture of the purity of the sample, and whether the ensemble is composed of mostly metallic or mostly semiconducting nanotubes. In the same vein, it can also probe the purity and electronic character of an isolated CNT.

As the reader might have guessed from the preceding discussion, AFM and SEM are somewhat complementary techniques that provide similar information about nanotubes. The key difference is that SEM is much faster in scanning a surface for nanotubes. However, owing to charging or other effects that can be present in an electron microscope, it is not as reliable a method for the accurate measurement of the diameter of single-wall nanotubes (say <5 nm). These CNTs are more accurately measured using an atomic force microscope to probe their heights. As a result, rapid determination of the diameter distribution of individual nanotubes is best achieved with AFM. For utmost measurement accuracy, a TEM can be employed.

Advanced techniques are also utilized to acquire further information about the nanotube electro-physical properties. Transmission electron microscopes are particularly useful to determine the number of shells present in a multi-wall CNT, and to check visually for structural defects. A high-resolution scanning tunneling microscope (STM) can actually render a visual image of a single-wall nanotube with atomic resolution, affording a precise determination of the nanotube chirality. Chirality is the only unique characteristic of a nanotube that determines its exact diameter and enables the most accurate computation of all its solid-state properties. Chirality will be discussed in detail in Chapter 4. It is worthwhile noting that an STM was employed to render the first image of the carbon–carbon bonding arrangement of a single-wall nanotube in agreement with the illustrative model (Figure 1.1a).[27] Optical spectroscopy is a more routine (but less accurate) alternative method of determining the chirality of single-wall nanotubes,[28] and X-ray photoelectron spectroscopy (XPS) is a useful (but not very flexible) method of probing the substrate surface before, during, and after the growth of nanotubes in order to elucidate the surface science and growth mechanisms.[29] Readers interested in any of these techniques will find a wealth of information available on the Internet regarding the theory, capabilities, inherent limitations, and manufacturers of these characterization systems.

1.5   What about non-CNTs?

There are several other lesser known nanomaterials that can be synthesized using a template procedure to form hollow tubular structures with morphology similar to CNTs. These include hollow tubes made from the oxide of various elements, such as vanadium, titanium, iron, copper, and aluminum, as well as oxides of

---

[27] J. W. G. Wilder, L. C. Veneng, A. G. Ringler, R. E. Snalley and C. Dekker, Electronic structure of atomically resolved carbon nanotubes. *Nature*, **391** (1998) 59–62.

[28] S. Bachilo, M. S. Strano, C. Kittrell, R. H. Hauge, R. E. Smalley and R. B. Weisman, Structure-assigned optical spectra of single-walled carbon nanotubes. *Science*, **298** (2002) 2361–6.

[29] D. Akinwande, N. Patil, A. Lin, Y. Nishi and H.-S. P. Wong, Surface science of catalyst dynamics for aligned carbon nanotube synthesis on a full-scale quartz wafer. *J. Phys. Chem. C*, **113** (2009) 8602–8.

semiconductors.[30] Similarly, nanotubes from pure metals such as copper have been reported,[31] while others from pure semiconductors such as silicon have yet to be observed but have been studied theoretically and predicted to exist under optimum conditions.[32]

The closest tubular nanomaterials to CNTs and next in importance are boron nitride nanotubes (BNTs). Inorganic BNTs have a bonding arrangement identical to CNTs, but with the carbon atoms alternately replaced by boron and nitrogen atoms. Just like CNTs, they can be synthesized to be either the single-wall or multi-wall flavor, though the synthesis methods have to be optimized accordingly. While BNTs share many similar physical properties to CNTs, their electronic or optical properties are, however, markedly different. Carbon nanotubes can be semiconducting or metallic depending on the chirality and diameter, whereas BNTs are insulators or wide bandgap semiconductors with a bandgap of $\sim$5.5 eV that is largely independent of tube chirality for practical diameters ($\gtrsim$1 nm). Moreover, CNTs offer direct optical transitions, whereas BNTs can offer either direct or indirect optical transitions, depending on the chirality.

For further studies on BNTs, the reader will find the comprehensive review by Golberg *et al.* particularly enlightening.[33]

[30] For examples, see: F. Krumeich, H.-J. Muhr, M. Niederberger, F. Bieri, B. Schnyder and R. Nagper, Morphology and topochemical reactions of novel vanadium oxide nanotubes, *J. Am. Chem. Soc.*, **121** (1999) 8324–31; and J. Huang, W. K. Chim, S. Wang, S. Y. Chian and L. M. Wang, From germanium nanowires to germanium–silicon oxide nanotubes: influence of germanium tetraiodide precursor. *Nano Lett.*, **9** (2009) 553–9.

[31] M. V. Kamalakar and A. K. Raychaudhuri, A novel method of synthesis of dense arrays of aligned single crystalline copper nanotubes using electrodeposition in the presence of a rotating electric field. *Adv. Mater.*, **20** (2008) 149–54.

[32] R. Q. Zhang, H.-L. Lee, W.-K. Li and B. K. Teo, Investigation of possible structures of silicon nanotubes via density-functional tight-binding molecular dynamics simulations and *ab initio* calculations. *J. Phys. Chem. B*, **109** (2005) 8605–12.

[33] D. Golberg, Y. Bando, C. C. Tang and C. Y. Zhi, Boron nitride nanotubes. *Adv. Mater.*, **19** (2007) 2413–32.

2 Electrons in solids: a basic introduction

Free your mind.

2.1 Introduction

The central purpose of this book is to understand the properties of electrons in CNTs and graphene. A good understanding is of utmost importance because it enables us to make electronic devices and engineer the performance of the devices to satisfy our desires. These devices can include, for example, sensors, diodes, transistors, transmission lines, antennas, and electron emission devices. In addition, the devices made out of carbon nanomaterials are being considered as building blocks for future applications broadly referred to as nanoelectronics, which includes circuits and systems. The technology to make nanomaterials and related devices is called nanotechnology.

To accomplish our central purpose, it is essential that we are familiar with the mathematical techniques and physical ideas behind the theory of electrons, particularly in solids. Specifically, in order to understand and describe the behavior of electrons in a solid requires consideration of:

 (i) The general quantum mechanical wave nature of electrons.
(ii) The periodic arrangement of atoms in crystalline solid matter, which is frequently called the crystal structure or lattice.

The introductory discussion of electrons in solids in this chapter will proceed in a manner that is beneficial for developing intuition, by considering an introductory quantum mechanical description of electrons, and subsequently exploring the crystal structure. Our attention throughout the chapter will be focused on the mathematical techniques, central ideas, and main results regarding electrons in solids. In view of the fact that quantum mechanics and solid-state physics are themselves fundamental disciplines of physics of great breadth and depth, this chapter is primarily intended as an elementary review of the *minimum* basic concepts and techniques in the quantum mechanical description of electrons in solids. The understanding and intuition developed in this chapter will be applied throughout the entire book.

References will be provided along the way for readers that would enjoy a more detailed coverage of the basic concepts. Readers already familiar with the physics concepts and techniques are invited to advance to the next chapter.

2.2   Quantum mechanics of electrons in solids

Understanding how electrons behave in a solid is central to the development of any solid-state electronic device. This understanding is what makes it possible to control electrons by external forces and to obtain interesting characteristics. To guide our intuition we will first review how electrons behave in free space and how their behavior changes when they are in a solid. By behavior we are mostly interested in their most fundamental properties, which are their wavevector and energy. Together these two properties give us an idea of how free or excited the electrons are behaving. The more excited they are, the easier it will be to control them. Energy and wavevector are actually related, and this relationship is commonly called the dispersion or band structure. It is not an overstatement to assert that the *dispersion is the most important and central characteristic that describes the behavior of electrons in a crystalline solid*. Indeed, it is the dispersion or band structure that we seek to derive and understand in this chapter, and subsequently in Chapters 3 and 4.

It is worthwhile noting that there are a gazillion[1] electrons present in typical solids. For example, there exist on the order of 10^{22} electrons/cm^3 if we consider only the valence electrons in bulk metals. How can we accurately describe the behavior of all these electrons in a solid? The simple answer is that we cannot describe them all, at least not exactly. However, if the electrons have negligible or no interaction with each other, than the problem reduces to the case of describing the behavior of one electron, which is by far a much simpler problem. This is technically called the *one-electron* or *independent electron* approximation. Fortunately, this approximation is accurate in understanding the behavior of electrons in the majority of solids of interest operating at room temperature. The independent electron is an underlying approximation employed in all the electron models developed in this chapter and in the entire textbook, unless noted otherwise. Figure 2.1 illustrates several one electron models in order of increasing complexity. This chapter will examine the first three models in the figure, and Chapter 3 will explore the tight-binding model for the development of the band structure of graphene.

Since the electrons will be treated as waves, Schrödinger's equation in its most basic form will be employed to solve for their properties. For this purpose, we assume the reader has at least a basic exposure to quantum mechanics at the level

---

[1] Think of a very, *very* large number whose precise value is unnecessary to specify at this moment.

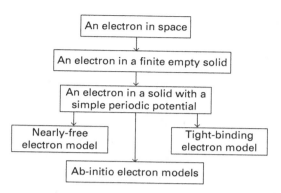

Fig. 2.1   Illustrative flowchart of *independent electron* models. Arrows indicate increasing complexity.

of an undergraduate modern physics course.[2] In general, the quantum mechanics in this section and in the entire book is kept to a minimum, and much of the understanding and intuition developed as we move ahead should still be accessible to readers who are not familiar with quantum mechanics.

2.3   An electron in empty space

The simplest possible model of electrons is the model of a free electron in empty space. Understanding this model is the first step towards our goal of gaining deeper insight and understanding of electrons in real crystalline solids. In principle, empty space is essentially the limiting case of an infinite force-free solid. *Free electrons* are electrons that have no forces or potential acting on them and hence, they are *free*. Even though we are considering a single free electron in these introductory models, the mathematical formalism developed and knowledge acquired applies directly to realistic solids with a much larger number of electrons, provided the electrons have negligible interaction with one another, which is often the case at room temperature.

In this simple pedagogical model of a free electron in space we seek to determine the energy–wavevector relationship or dispersion by solving the time-independent[3] Schrödinger equation. In one-dimensional (1D) space measured by

---

[2] A widely read modern physics textbook is D. Giancoli, *Physics for Scientists and Engineers with Modern Physics*, 4th edn (Prentice Hall, 2008). For introductory graduate-level coverage of quantum mechanics see D. A. B. Miller, *Quantum Mechanics for Scientists and Engineers* (Cambridge University Press, 2008).

[3] The *time-independence* terminology can be quite confusing to the new aspiring quantum mechanic by suggesting there is no time dependence, which is certainly not the case for a wave. In the language of quantum mechanics, what is meant is that the probabilistic properties of the system are *time invariant* (to borrow jargon from signals theory).

the variable x, the Schrödinger equation is

$$H\psi(k,x) = E(k)\psi(k,x) \tag{2.1}$$

$$\left(-\frac{\hbar^2}{2m}\frac{\partial^2}{\partial x^2} + U(x)\right)\psi(k,x) = E(k)\psi(k,x), \tag{2.2}$$

where H is the Hamiltonian operator with its explicit form in the parentheses of Eq. (2.2),[4] ψ is the electron wavefunction, E is the electron total energy, and k is the wavevector. $\hbar$ is the reduced Planck's constant, m is the electron mass, and $U(x) = 0$ for all x for the free-electron model. This is a frequently occurring differential equation with a general solution in the form of exponentials:

$$\psi(k,x) = Ae^{ikx}, \tag{2.3}$$

where k can take on arbitrary real positive or negative values and where the complex exponential with an amplitude A is considered a plane wave because it oscillates similar to electromagnetic waves in space. This is easily seen by expanding the exponential in terms of trigonometric functions. To determine the energy of the electrons we employ two relations. The first relation from classical mechanics treats the electron as a particle with an energy given in terms of its momentum:

$$E = \frac{p^2}{2m}, \tag{2.4}$$

where p is the momentum of the particle. The second relation is the de Broglie wave-duality postulate, which assigns wave-like behavior (vis-à-vis k) to a particle with momentum p, $p = \hbar k$.[5] Substituting $p = \hbar k$ into Eq. (2.4), the energy–wavevector relationship for a free electron in space leads to the celebrated parabolic dispersion:

$$E = \frac{\hbar^2 k^2}{2m}. \tag{2.5}$$

For three-dimensional (3D) space, it is a straightforward exercise to show that the dispersion is essentially the same with the wavevector k replaced by the 3D wavevector $\mathbf{K}$, and $\mathbf{K}^2 = k_x^2 + k_y^2 + k_z^2$.

In summary, the key observations we gain by examining free electrons in space are that the electrons oscillate as plane waves with a continuous set of wavevectors and, correspondingly, a continuous energy spectrum. The study of this pedagogical

---

[4] The Hamiltonian operator acts on the wavefunction to produce the corresponding allowed energies.

[5] As it is said "success has many fathers." Louis de Broglie's postulate of matter waves (his Ph.D. thesis) was a generalization of Albert Einstein's energy quanta theory of the photoelectric effect, which itself builds on Max Planck's blackbody quantization theory. All three were awarded (on separate occasions) a Nobel Prize for their development of quanta and quantum mechanics.

model by itself is not particularly interesting; however, when we compare it with the behavior of electrons in a finite solid we begin to develop intuition as to how the solid environment perturbs the electrons and the general technique of solving for the energy–wavevector dispersion.

2.4 An electron in a finite empty solid

We consider the next simplest model of electrons by reducing the empty *infinite* solid of the previous section to an empty *finite* solid of length L with a potential illustrated in Figure 2.2 and mathematically described in one dimension as

$$U(x) = \begin{cases} 0 & \text{for } 0 < x < L \\ \infty & \text{elsewhere.} \end{cases} \tag{2.6}$$

How the dispersion of electrons in this finite solid compares with the previous model is the primary question of interest. This idealized model is also called by phrases such as *particle in a box or particle in an infinite potential well*. However, a concern to keep in mind is that finite empty solids with infinite potential walls do not exist in nature. Nonetheless, the finite empty solid model is of great technical value because it shines light on our intuition about electrons in real solids and it provides a basis for developing simple analytical dispersions of far more complicated crystalline solids. In principle, this model is essentially the limiting case of a solid in which the electrostatic force between the atomic nucleus and the electrons under consideration is negligible.

   Having convinced ourselves that this model is of value in developing insight, we proceed with the analysis by solving Schrödinger's equation (Eq. (2.2)) with the potential given by Eq. (2.6). The general solution is given by complex exponential functions indicative of an oscillating wave:

$$\psi(k, x) = Ae^{ikx} + Be^{-ikx}. \tag{2.7}$$

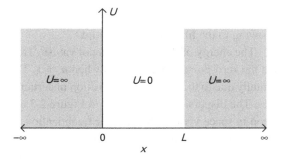

Fig. 2.2   The potential profile that defines a 1D finite *empty* solid of length L.

The finite size of the solid imposes a boundary condition that places a restriction on the wavefunction.

$$\psi(x = 0) = \psi(x = L) = 0. \tag{2.8}$$

A hand-waving argument to justify these boundary conditions is that the electron cannot escape from the solid since the infinite potential exerts an infinite force that bounds the electron. As a consequence, the wavefunction and the probability of finding the electron must vanish outside the solid. Applying the boundary condition at $x = 0$ requires $B = -A$, and simplifies the wavefunction, $\psi(k, x) = C \sin kx$, where C is generally a complex constant to be determined shortly. The boundary condition at $x = L$ requires

$$k = \frac{n\pi}{L}, \quad n = 1, 2, 3, \ldots, \tag{2.9}$$

where we have ruled out $n = 0$ because, for that case, $\psi = 0$ everywhere, implying that the electron is not present in all space, which is a trivial case of no interest. n is called a quantum number that indexes the allowed wavevectors. Substituting for the wavevector into Eq. (2.5), the resulting energies are

$$E = \frac{h^2 k^2}{2m} = \frac{h^2 \pi^2}{2mL^2} n^2. \tag{2.10}$$

The only remaining unknown is the amplitude C, which is determined from a normalization condition, affirming that the probability of finding the electron in the box is unity:

$$\int_{-\infty}^{\infty} |\psi(x)|^2 dx = \int_0^L |\psi(x)|^2 dx = 1, \tag{2.11}$$

where $|\psi(x)|^2$ is interpreted as the probability density. The normalization condition results in $|C| = \sqrt{2/L}$, or simply $C = \sqrt{2/L}$, if we consider only real-valued numbers. The important physics gained from this analysis is as follows:

1. *Dispersion quantization*   The electron wavevector and energy can only take on discrete values owing to the finite size of the solid.
2. *Zero-point energy*   The energy of the electron cannot vanish. The lowest possible energy, called the zero-point energy, occurs when $n = 1$. This minimum energy is fundamentally due to the momentum–position uncertainty relation in quantum mechanics. The dispersion is illustrated in Figure 2.3 and compared with the free electron in space vividly showing the quantization due to spatial confinement. It is a straightforward exercise to show that the dispersion is readily extended to a 3D empty finite solid by replacing the wavevector k with the 3D wavevector $\mathbf{K}$, which leads to three independent quantum numbers, $n_x, n_y,$

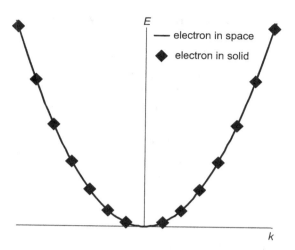

Fig. 2.3   Energy–wavevector dispersions of a free electron in space and in a finite solid.

and n_z representing the boundary conditions from each spatial dimension. The resulting 3D dispersion is

$$E = \frac{\hbar^2 \pi^2}{2mL^2}(n_x^2 + n_y^2 + n_z^2). \tag{2.12}$$

Compared with the 1D dispersion, the 3D dispersion contains the same physics plus an *additional physics* called *degeneracy*, which is defined as the case where a different set of quantum numbers results in the same energy. For example, the set of quantum numbers

$$(n_x, n_y, n_z) = (1, 2, 3); (1, 3, 2); (3, 1, 2);$$
$$= (2, 1, 3); (2, 3, 1); (3, 2, 1)$$

result in the same energy and, therefore, are sixfold degenerate. In general, if m sets of quantum numbers yield the same energy, this is called an m-fold degeneracy. Degeneracy is an important concept, as we shall see throughout this book, and often reflects the underlying geometrical symmetries[6] present in the system.

The utility of the simple parabolic dispersion obtained here cannot be overemphasized. It has been used successfully (with few modifications) to predict and explain many of the properties of electrons from 3D solids such as bulk semiconductors to 1D molecular solids such as CNTs.

---

[6] Symmetry is a fundamental idea in quantum mechanics and solid-state physics and is of great value in providing insight about how electrons conduct themselves in crystalline solids. Useful (and compact) discussions can be found in H. Kroemer, *Quantum Mechanics* (Prentice Hall, 1994); and N. W. Ashcroft and N. D. Mermin, *Solid State Physics* (Brooks/Cole, 1976).

2.5   An electron in a periodic solid: Kronig–Penney model

So far we have concerned ourselves with the behavior of an electron in an infinite and finite *empty* solid and become acquainted with the general nature of the dispersion and energy quantization that comes about because of size confinement. However, real crystalline solids have a periodic arrangement of atoms that exerts a force on electrons. Therefore, in reality, any electron in a solid cannot be completely free. In the pursuit of spatial freedom, an electron can, at best, be *nearly free*. The paramount question now is: How does the periodic potential of the atoms affect the behavior of an electron in a solid?

To address this question, we have to consider how to model the periodic potential field in the solid or crystal lattice as it is sometimes called. The actual potential field can be determined from Coulomb's law; however, for simplicity we shall consider periodic delta potentials in a large 1D solid lattice with length L, as shown in Figure 2.4, originally proposed by *Kronig and Penney*.[7] This simplicity is fortunately mostly mathematical in nature while gracing us with insight into much of the physics of electrons in real solids.

$$\frac{d^2\psi}{dx^2} + \frac{2m}{\hbar^2}(E - U(x))\psi = 0. \tag{2.13}$$

This is a well-studied differential equation with formal solutions that are described by Bloch's theorem in solid-state physics or equivalently by the much older Floquet's theorem of differential equations. These solutions are of the form

$$\psi(x) = u(x)\,e^{ikx}, \tag{2.14}$$

where $u(x)$ is an amplitude modulating function with the periodicity of the lattice potential, $u(x) = u(x+a)$. Bloch's theorem states that the wavefunction is periodic with the length of the lattice, i.e. $\psi(0) = \psi(L)$, which results in quantization of the wavevector (as we have observed in the previous empty solid model):

$$e^{ikL} = 1, \quad \rightarrow k = \frac{2\pi}{L}n, \tag{2.15}$$

where n is an integer. To determine $u(x)$ we consider one unit cell and examine the space between the potentials when $U(x) = 0$, whence Schrödinger's equation reduces to

$$\frac{d^2\psi}{dx^2} + \gamma^2\psi = 0, \quad \gamma^2 = \frac{2mE}{\hbar^2}. \tag{2.16}$$

---

[7] R. de L. Kronig and W. G. Penney, Quantum mechanics of electrons in crystal lattices. *Proc. R. Soc. (London) A*, **130** (1931) 499–513.

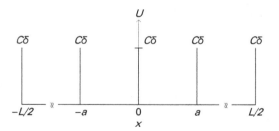

Fig. 2.4 Kronig–Penney periodic potential in a solid of length L. C is the strength of the delta function. Inside the solid, the electron wavefunction must satisfy Schrödinger's equation with a periodic potential $U(x) = U(x + a)$, and the length of the solid is an integral multiple of the unit cell.

Substituting Eq. (2.14) into Eq. (2.16) yields

$$\frac{d^2u}{dx^2} + 2ik\frac{du}{dx} + (\gamma^2 - k^2)u = 0. \tag{2.17}$$

This is a standard second-order linear differential equation with constant coefficients and has a solution given by

$$u(x) = (A \cos \gamma x + b \sin \gamma x)e^{-ikx} \tag{2.18}$$

subject to periodic boundary conditions that guarantee continuity for the $u(x)$ function and its slope:

$$u(0) = u(a). \tag{2.19}$$

The continuity conditions of the slope involving delta potentials have to be treated with care. We derive the appropriate boundary condition for the slope by integrating the Schrödinger equation over a very small region 2ϵ about $x = 0$:

$$-\frac{\hbar^2}{2m} \int_{-\epsilon}^{\epsilon} \frac{d^2\psi}{dx^2}dx + \int_{-\epsilon}^{\epsilon} C\delta(x)\psi dx = E \int_{-\epsilon}^{\epsilon} \psi dx. \tag{2.20}$$

In the right-hand side (RHS) of the equation, ψ does not change appreciably if the interval is infinitesimal and can be assumed constant, $\psi(0)$. Evaluating the integrals gives

$$\psi'(\epsilon^+) - \psi'(\epsilon^-) - \frac{2m}{\hbar^2}C\psi(0) = 2\epsilon E\psi(0). \tag{2.21}$$

In the limit that $\epsilon \to 0$, then $2\epsilon E\psi(0) \to 0$, the equation simplifies to

$$\psi'(\epsilon^+) - \psi'(\epsilon^-) = \frac{2mC}{\hbar^2}\psi(0). \tag{2.22}$$

Substituting for ψ (and its derivative) from Eq. (2.14) and simplifying yields

$$u'(\epsilon^+) - u'(\epsilon^-) + ik(u(\epsilon^+) - u(\epsilon^-)) = \frac{2mC}{\hbar^2}u(0). \qquad (2.23)$$

Without any loss in accuracy, we can replace $x = \epsilon^+$ with $x = 0$ and replace $x = \epsilon^-$ with $x = a$ since $u(x)$ is periodic with length a:

$$u'(0) = u'(a) + \frac{2mC}{\hbar^2}u(0). \qquad (2.24)$$

Finally, substituting Eq. (2.18) into Eq. (2.19), and also into Eq. (2.24), results in two simultaneous equations:

$$A(1 - e^{-ika} \cos \gamma a) = Be^{-ika} \sin \gamma a$$

$$A\left(ike^{-ika} \cos \gamma a + \gamma e^{-ika} \sin \gamma a - ik - \frac{2m}{\hbar^2}C\right)$$

$$= B(\gamma e^{-ika} \cos \gamma a - ike^{-ika} \sin \gamma a - \gamma),$$

which can be solved by the method of substitution, resulting in the following solution:

$$\cos ka = \cos \gamma a + P\frac{\sin \gamma a}{\gamma a}, \qquad (2.25)$$

where $P = maC/\hbar^2$. The solution is a transcendental equation that is not solvable algebraically. One often has to resort to graphical or numerical techniques. In the limit of zero potential ($P = 0$), the solution reduces to the case of a free electron in a finite solid.

There are several ways to digest the solution. Let us start by graphing the RHS of Eq. (2.25) as a function of γa as is shown in Figure 2.5 ($P = 5$ for this example). The RHS results in an oscillating function that is sometimes greater than $+1$ and sometimes less than -1. Comparing this with the left-hand side (LHS) of Eq. (2.25), which is restricted to values within ± 1, we can conclude that only certain values of γa (and hence energy via Eq. (2.16)) that satisfy this restriction result in allowed solutions of Eq. (2.25), and values that do not satisfy the restriction are forbidden. As Kronig and Penney eloquently stated in their 1931 article:

the energy values which an electron moving through the lattice may have, hence form a spectrum consisting of continuous pieces separated by finite intervals.

In other words, there are allowed energy bands separated by gaps, known as bandgaps. To see this more vividly, the dispersion[8] is shown in Figure 2.6, which

---

[8] One way to compute the dispersion numerically is to obtain the set of values of γa and, hence, the energies that satisfy the ± 1 restriction, as shown in Figure 2.5. This set of values organizes in

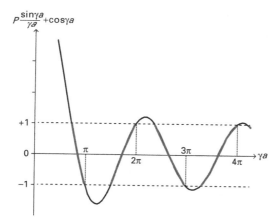

Fig. 2.5   Graph of the RHS of Eq. (2.25). The gray lines between ± 1 dashed lines are the allowed solutions. Only the positive γa is shown here; the negative γa is the mirror image.

clearly reveals a set of allowed energy bands separated by bandgaps. For comparison, the dispersion of a free electron in space has also been included in the figure.

2.6   Important insights from the Kronig–Penney model

It is worthwhile taking this opportunity to explain the other key features of the dispersion in Figure 2.6, beyond the discussion of allowed bands separated by bandgaps.

→ **Brillouin Zone**: Eq. (2.25) does not have a *unique* solution. For every allowed energy that yields an LHS value within ± 1 there will be correspondingly an infinite set of wavevectors k. For example, let us say a specific energy value leads to an LHS of $+1$. As a result, the wavevector calculated from $\cos ka = +1$ yields infinite acceptable k-values due to the repeating nature of the trigonometric functions. It is often convenient to obtain a finite and unique (non-repeating) set of allowed $E-k$ pairs. To achieve this, we consider wavevectors within a single period of the LHS. Additionally, we chose this period to be symmetric around the origin, i.e. $ka = [-\pi, +\pi]$. The space or zone restricted to these k-values is formally called the first Brillouin zone and is illustrated as the shaded region in Figure 2.6. Similarly, for every allowed wavevector there are infinite allowed energies, corresponding to one allowed energy per band, for a total of infinite bands. To see this visually, consider $\cos ka = 0$, which corresponds to the horizontal

bands demarcated by $\gamma a = \pi$. For example, the energies of the nth band satisfy the restriction lieing within $\gamma a = [(n-1)\pi, n\pi]$. Once the energies are known, the wavevector is computed from Eq. (2.25).

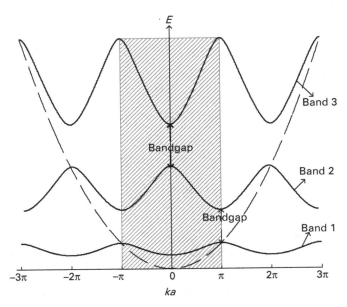

Fig. 2.6 The dispersion of an electron in the Kronig–Penney model showing allowed bands separated by bandgaps. The dashed curve is the dispersion of a free electron.

intersection of the curve passing through the origin (Figure 2.5). Owing to the oscillatory nature of the curve resulting from $\cos \gamma a$ in Eq. (2.25) there will be an infinite number of allowed energies for that given k-value. However, there are a finite number of electrons in any solid, and we will discover throughout this book (see Chapters 3 and 4 for example) that our focus will be on the properties of the highest energy electrons which are largely responsible for the device physics. How the highest energies of relevance are determined will be elucidated with the idea of the Fermi energy discussed in Chapter 3.

→ **Nearly-free Electron**: In the limit that $P \to 0$, then $k \to \lambda$ (see Eq. (2.25)) and we recover the dispersion for an electron in an empty solid of size L:

$$E = \frac{\hbar^2 k^2}{2m} = \frac{\hbar^2 \pi^2}{2mL^2} n^2, \quad n = \pm1, \pm2, \pm3, \ldots \tag{2.26}$$

This informs us that, in the limit of a weak periodic potential, the electron would increasingly behave as if it were nearly free; that is, *it can move around in the solid with negligible or no impediment from the periodic potential.*[9] Indeed, there is a widely used model in solid-state physics called the *nearly-free electron* model that assumes the electron behaves like a free electron but it only slightly perturbed by the periodic potential, which is essentially accounted for (to first order) as a

---

[9] This can be considered the most accurate description of what we mean by a "free (or nearly-free) electron" model.

constant potential in the Schrödinger equation (Eq. (2.1)). The nearly free electron model is particularly useful in predicting fairly accurate band structures of bulk metals and the existence (or lack thereof) of bandgaps in crystalline solids.

→ **Tightly Bound Electron**: In the opposite limit, when $P \to \infty$, then the solutions to Eq. (2.25) exist only when $\gamma a = n\pi$, and the dispersion becomes

$$E = \frac{\hbar^2 k^2}{2m} = \frac{\hbar^2 \pi^2}{2ma^2} n^2, \quad n = \pm 1, \pm 2, \pm 3, \ldots, \tag{2.27}$$

which at first glance appears similar to Eq. (2.26) with the important difference that Eq. (2.27) corresponds to the case of an electron moving within a space of size a. This is interpreted as an electron *trapped* or *localized* within the infinite potential walls; as a result, *it is not free to move around the entire solid of size L,* in contrast to the nearly-free electron case. This informs us that, in the limit of an increasingly strong periodic potential, the electron behaves as if it were *tightly bound* by the strong potential, which in reality represents the attraction from the atomic nucleus. The case of a strong periodic potential is of significant practical interest and is formally treated by methods such as the *tight-binding model*. In fact, this is the most widely used model for accurately describing the band structure of graphene and carbon nanotubes. This model will be discussed in more detail in the derivation of the band structure of graphene.

→ **Parabolic-like Dispersions**: We learned from the Kronig–Penney model that the dispersion is fairly parabolic in structure, particularly at the bottom or top of each band (see Figure 2.6). While we utilized a delta periodic potential in the specific model, the qualitative features of the band structure are fairly insensitive to the specific form of the periodic potential. A case in point is a repeating square potential which also yields a parabolic-like band structure. To a great degree, many (but not all) solids, including metals and semiconductors, exhibit parabolic-like dispersions at the bottom and top of their bands of interest. As a result, there has been great intellectual activity in developing simple algebraic equations based on modifications of the free-electron parabola to describe the dispersion (and resulting device physics) of solids, particularly semiconductors, with varying degrees of accuracy. The algebraic expression generally takes one of the following analytical forms:

$$E \approx \frac{\hbar^2 k^2}{2m_{\text{eff}}} \tag{2.28}$$

$$E(1 + \alpha E) \approx \frac{\hbar^2 k^2}{2m_{\text{eff}}}, \tag{2.29}$$

where m_{eff} represents the effective mass of electrons specific to the particular solid and conceptually embodies the effect of the periodic potential on the motion or mobility of the electrons in the solid. The non-parabolicity parameter α quantifies how the actual dispersion of interest compares with an ideal free-electron parabola and is (approximately) inversely proportional to the bandgap. These three parameters are often determined in practice by fitting the algebraic equation to experimental data or extracted directly from *ab-initio*[10] computation.

We have discussed the Kronig–Penney model of electrons in a periodic potential quite extensively; on the other hand, we have yet to address why bandgaps exist in the first place. It is not at all instantly obvious why bandgaps should exist. However, when we view the electrons strictly as waves, then wave principles, such as reflections, are applicable. It is the reflections of electron waves from the potential wall that lead to bandgaps. And, analogous to electromagnetic waves, these reflections are an integral multiple of wavelength, and that is why we observe the bandgaps occurring at $k\alpha = \pm n\pi$. In solid-state physics, these reflections are called *Bragg reflections*.

2.7   Basic crystal structure of solids

In the previous section we gained much insight into how electrons behave in a crystalline solid (a solid with a periodic potential). The Kronig–Penney model, for instance, is not only of great educational value, but its dispersion is often also used to fit the actual dispersion of many bulk solids. Nonetheless, the Kronig–Penney model is inherently limited because it is a general model that does not take into account the specific lattice arrangement of a particular solid of interest. For example, if we desire to determine the band structure of nanotubes and nanoribbons, the Kronig–Penney model is utterly inadequate in providing basic information such as the bandgap dependency on the diameter of these nanostructures.

In order to make any further progress in determining the *specific* band structure of a particular solid, we would need to understand the crystal structure of that particular solid. *To underscore the importance of the crystal structure, it would be fair to state that it is at the core of modern solid-state physics.* The crystal structure can be viewed in two domains similar to how a coin has two views. One domain is in position (sometimes called direct) space and the other domain is in reciprocal space. We will start by discussing the lattice arrangement in position space, subsequently followed by a corresponding discussion in reciprocal space. Our focus here will be on discussion of the crystal structure that is relevant to this text, namely for the development of two-dimensional (2D) or reduced-dimensional solids (graphene and CNTs). As such, the spotlight will be on lattices in 2D space. For a more general coverage of crystal structures, the reader will find a general solid-state textbook very informative.[11]

---

[10] *Ab initio* is a Latin term that means from *first principles* or *from the beginning*. In science, it refers to a thorough calculation that starts directly from accepted laws (such as Schrondinger's equation) and does not use empirical data or fitting in the calculation.

[11] For an introductory graduate-level coverage see C. Kittel, *Introduction to Solid State Physics*, 8th edition (Wiley, 2004). For more advanced studies, consider the classic text N. W. Ashcroft and N. D. Mermin, *Solid State Physics* (Brooks/Cole, 1976).

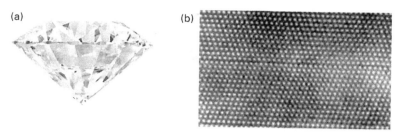

Fig. 2.7 (a) Macroscopic image of a "round brilliant cut" diamond (courtesy europeanj.com). (b) High-resolution TEM image of the crystal lattice of diamond. Inter-atomic distance is ~2–2.1 Å. Copyright (1990) Materials Research Society. Courtesy of B.E. Williams, H.S. Kong, and J.T. Glass, *J. Mater. Res.*, **5** (1990) 801.

Figure 2.7a shows the macroscopic appearance of a diamond crystal with its visible symmetric crystal planes. However, the physical glamour of a solid is not a sufficient measure for crystallinity. What is required is microscopic periodicity of the lattice. Figure 2.7b is a high-resolution TEM image of a region in a diamond crystal with noticeable periodicity, where each bright spot represents a carbon atom. Our concern will be on the characteristic underlying microscopic lattice of the crystal structure. As such, a systematic mathematical formalism is needed to describe the geometry of the lattice. From this mathematical formalism we can then develop a technique to compute the band structure of particular solids. To foreshadow the discussion to come, it is worthwhile stating in advance the logical relation which is the foundation of crystal structure,

$$\text{Crystal Lattice} = \text{Bravais Lattice} + \text{Basis}$$

where Bravais lattice is the technical name for the fundamental irreducible microscopic arrangement of points and the basis is an integer number that represents how many atoms are attached to each Bravais lattice point.

2.8 The Bravais lattice

There are several ways to describe a lattice. The most fundamental description is known as the Bravais[12] lattice. In words, a Bravais lattice is an array of discrete points with an arrangement and orientation that look exactly the same from any of the discrete points; that is, the lattice points are indistinguishable from one another. The Bravais lattice point itself represents some "stuff," where "stuff" could be a single atom, a collection of atoms, molecules, and so on and so forth. The amount

[12] Named after the French physicist Auguste Bravais who correctly pointed out in the mid 1800s that there are 14 unique fundamental lattices in 3D space, revising the earlier calculation of 15 lattices by Frankenheim, otherwise they would be known as Frankenheim lattices today.

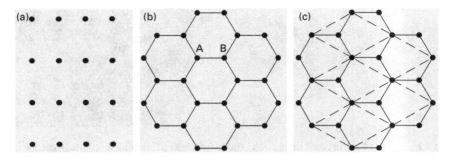

Fig. 2.8   (a) A rectangular lattice qualifies as a Bravais lattice. (b) A honeycomb lattice is not a Bravais lattice. (c) Conversion of the honeycomb into a Bravais lattice with a basis of 2. The underlying Bravais lattice is represented by dashed lines.

of "stuff" at each Bravais lattice point is what we earlier called the basis. It is significantly easier to grasp the meaning of a Bravais lattice and basis visually. Let us consider the lattices shown in Figure 2.8. Every discrete point in the 2D rectangular lattice sees exactly the same environment. Imagine placing a camera at one of the discrete points and taking a picture to the right of the point. That picture will look exactly the same if it were taken from any other discrete point in the lattice. As a result, the rectangular lattice is a Bravais lattice and has a basis of one. That is, every Bravais lattice point represents a single crystal lattice point.

Let us perform the same thought experiment with the camera on the honeycomb lattice in Figure 2.8b. Let us say that initially the camera is at point A and the image is taking facing rightwards. If we move the camera to point B facing rightwards, then the resulting image from B is obviously not the same as the image from A. For one, point A is directly looking at a single lattice point, while point B faces two lattice points at angles of $\pm 60°$. Consequently, the honeycomb, *as it is*, is not a Bravais or fundamental lattice. However, the honeycomb crystal *is* still a lattice because it consists of repeating unit cells and, as a result, it can be converted or mapped to a Bravais lattice. The process of converting a non-fundamental lattice to a Bravais lattice occurs most conveniently by literally staring at the lattice (long enough) to identify a collection of crystal lattice points that when grouped together will result in a Bravais lattice. For the particular example of a honeycomb, consider grouping points A and B together (and all repeating instances of A and B); the resulting lattice is shown in Figure 2.8c, which is now a Bravais lattice, and has a basis of two. That is, every Bravais lattice point represents two crystal lattice points. In crystalline solids, the real points are locations of atoms and the basis is conveniently interpreted as the number of atoms grouped together at each Bravais lattice point. It is important to keep in mind that the Bravais lattice is not always the same as the crystal lattice, as this example demonstrates.

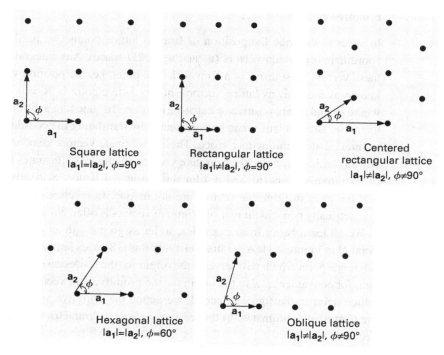

Fig. 2.9 The five unique Bravais lattices in 2D space.

There are precisely five Bravais lattices in 2D space (Figure 2.9) and 14 Bravais lattices in 3D space. These precise totals are determined from symmetry properties of Bravais lattice points. For instance, the square lattice is invariant under rotation of 90° about a fixed lattice point in 2D space. The hexagonal lattice, on the other hand, is invariant under rotation of 60°. By carefully taking into account the rotation symmetry, and symmetry present in lattice spacing, the Bravais lattices can be fully enumerated. The mastery of all the details of crystallographic symmetry[13] is (fortunately) not necessary for an advanced working knowledge of crystalline solids.

A useful property of crystal lattices which we will later employ in the derivation of electronic band structures is the nearest neighbors. The nearest neighbors are the number of points that are equally closest to a given point in the Bravais lattice. For example, the nearest neighbors are two and four for any point in the rectangular and square lattices respectively (see Figure 2.9).

---

[13] There is a tremendous amount of sophisticated detail formulated in the language of group theory, which is a discipline within *abstract algebra*. Further reading on group theory can be found in the references in footnote 6.

Primitive vectors

In order to describe the position of Bravais lattice points, we need to define a coordinate origin and vectors ($\mathbf{a}_1$ and $\mathbf{a}_2$ in 2D space). Any integral multiple of these vectors must arrive at a Bravais lattice point, i.e. the position vectors (also known as the Bravais lattice vectors) of any lattice point is $\mathbf{R} = n_1\mathbf{a}_1 + n_2\mathbf{a}_2$, where n_1 and n_2 are positive or negative integers. To state it in a different manner, the entire Bravais lattice can be constructed by translating the coordinate vectors in integral steps throughout space. These coordinate vectors associated with the Bravais lattice are called the primitive vectors. Figure 2.10 illustrates four choices of primitive vectors ($\mathbf{a}_1$ and $\mathbf{a}_2$) for the hexagonal lattice, evidently conveying that the pairs of primitive vectors are not unique. In practice, a symmetrical or geometrically convenient pair of primitive vectors is often chosen for simplicity.

We all learn better from examples, so let us go through an example to locate point Q in Figure 2.11. After first verifying it is a Bravais lattice, we then define a reference point which will serve as the origin for the Cartesian coordinate system and, for convenience, also the origin for the primitive unit vectors (the light gray lattice point in the figure). Second, we define the primitive vectors in terms of the Cartesian coordinates. In this case, we chose a symmetrical set of primitive vectors.

$$a_1 = \sqrt{3}\frac{a}{2}\hat{x} + \frac{a}{2}\hat{y}, \quad a_2 = \sqrt{3}\frac{a}{2}\hat{x} - \frac{a}{2}\hat{y}, \tag{2.30}$$

where a is the lattice constant, and for the hexagonal lattice $a = |\mathbf{a}_1| = |\mathbf{a}_2|$. Finally, it is a simple matter to locate point Q by traveling two primitive vectors in the negative $\mathbf{a}_2$ direction and one primitive vector in $\mathbf{a}_1$ (the dashed arrow lines in the figure). The Bravais lattice vector for point Q is $\mathbf{R} = \mathbf{a}_1 - 2\mathbf{a}_2$.

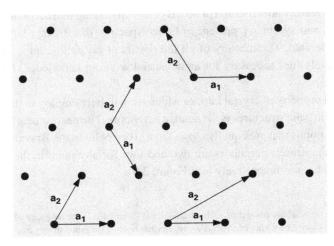

Fig. 2.10   Hexagonal Bravais lattice with several choices for pairs of primitive vectors.

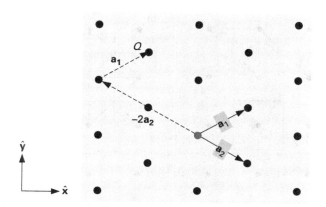

Fig. 2.11   An example using primitive vectors to determine the location of point Q in the lattice.

Primitive unit cell

The most basic unit cell of the Bravais lattice is called the primitive unit cell. The primitive unit cell has two distinguishing properties: (1) it is a region of space that contains exactly one Bravais lattice point; (2) it recreates the lattice when translated through all the Bravais lattice vectors without leaving gaps or generating overlaps. From the former property, the density of points in the primitive unit cell is one point/volume, where volume refers to the volume of the primitive cell. An additional observation from the former property is that the actual number of crystal lattice points contained in the primitive unit cell is equal to the basis. Similar to the primitive vectors, there is no unique way of choosing a primitive unit cell for a given Bravais lattice. Fortunately, in 2D space this is a relatively trivial matter, as it is easy to identify a primitive unit cell for any of the five Bravais lattices by visual inspection. Figure 2.12a illustrates some examples of primitive unit cells for the square lattice. All primitive unit cells occupy exactly the same area or volume.

In three dimensions, it can become painfully difficult to visualize and sketch Bravais lattices,[14] let alone construct a 3D primitive unit cell. To overcome this, a primitive unit cell known as the Wigner–Seitz cell has become broadly utilized, and is constructed from straightforward geometrical rules.[11] The Wigner–Seitz cell is especially very valuable in understanding the properties of the reciprocal lattice and the band structure which will be developed in the subsequent section and chapter.[15] Formally, the Wigner–Seitz cell is a primitive unit cell that has a boundary about a Bravais lattice point that contains only that lattice point. That boundary is closest to that lattice point than to any other lattice point. What does

---

[14] This is particularly true for the non-artists. More often than not, students who have a talent for sketching find themselves excelling in solid-state physics (at least in the areas of crystallography) compared with their artistically challenged peers.

[15] This is the primary incentive to learn the construction of the Wigner–Seitz cell in 2D space.

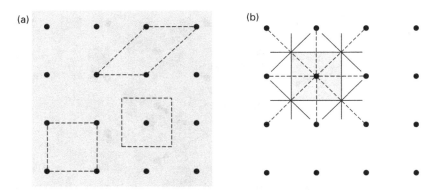

Fig. 2.12   Square lattice. (a) Several choices of primitive unit cells. (b) Construction of the
Wigner–Seitz cell (gray square).

that really mean? We find again that a picture is often more informative than many
words. The Wigner–Seitz cell for the square lattice is shown in Figure 2.12b. The
cell is constructed from the following recipe:

- Draw a connection line linking the lattice point to adjacent lattice points, often
  up to the second nearest neighbors. The connection lines are dashed lines in
  Figure 2.12b.
- Bisect the connection lines with perpendicular lines (solid lines in Figure 2.12b).
- Connect the bisecting lines at common intersections to form a closed polygon
  (shaded region in Figure 2.12b).

While the ensuing square Wigner–Seitz cell for the square lattice appears fairly
uninteresting, the inquisitive reader will find it enlightening and even enjoyable to
construct the Wigner–Seitz cell for the other 2D Bravais lattices. In those cases,
the Wigner–Seitz cell is always a hexagon except for the rectangular lattice. The
Wigner–Seitz cell exhibits the symmetry of the corresponding Bravais lattice; for
example, the rotational symmetry is invariant. To keep the focus on the main
subject of this text (2D solid matter), we will not discuss 3D Wigner–Seitz cells,
which can have very complex structures. Needless to say, such discussion can be
readily found in any standard texts on solid-state physics.

In this section, we have learned about four concepts which are important for
obtaining a working knowledge of crystalline solids. These concepts are Bravais
lattice, lattice basis, primitive unit vectors, and the primitive unit cell or Wigner–
Seitz cell.

2.9   The reciprocal lattice

In the prior section we elaborated on the crystal lattice. In general, it is composed
of a Bravais lattice and a basis. The Bravais lattice, which is normally measured

in terms of a position vector, is also known as the direct lattice. This distinction is necessary to maintain clarity and keeps the discussion tractable going forward.[16] The central feature of a lattice is periodicity or regularity. In the study of periodic order, Fourier analysis naturally comes to mind. A case in point is the study of periodic signals; one can view the signal in the time domain and also view the signal in the frequency domain by employing Fourier analysis. Our objective in this section is to apply Fourier analysis to the study of the lattice in order to gain additional insights. Without much ado, *the reciprocal lattice is the discrete Fourier transform of the direct lattice*.[17]

An elementary way to view the existence of a reciprocal lattice is to recall that numbers, functions, and matrices all have reciprocal equivalents, generally related by $[X^{-1}] \cdot [X] = 1$, where X can symbolize numbers, functions, and matrices and $[X^{-1}]$ is the reciprocal equivalent. Similarly, temporal space has frequency domain as its reciprocal; and the direct lattice (which is nothing more than a position space with periodicity) has the reciprocal lattice has its reciprocal equivalent. The reciprocal lattice satisfies the basic relation

$$e^{i\mathbf{K} \cdot \mathbf{R}} = 1, \tag{2.31}$$

where $\mathbf{K}$ is the set of wavevectors that determine the sites of the reciprocal lattice points and $\mathbf{R}$ is the Bravais lattice position vector as usual. The basic relation originates from the Fourier analysis of the direct lattice. The reciprocal lattice is a concept of paramount importance and provides fundamental insight into the behavior of electrons in crystalline solids. In three dimensions the primitive vectors of the reciprocal lattice ($\mathbf{b}_1, \mathbf{b}_2$, and $\mathbf{b}_3$) are determined from the primitive vectors of the direct lattice:

$$\mathbf{b}_1 = 2\pi \frac{\mathbf{a}_2 \times \mathbf{a}_3}{\mathbf{a}_1 \cdot (\mathbf{a}_2 \times \mathbf{a}_3)}, \qquad \mathbf{b}_2 = 2\pi \frac{\mathbf{a}_3 \times \mathbf{a}_1}{\mathbf{a}_1 \cdot (\mathbf{a}_2 \times \mathbf{a}_3)}, \qquad \mathbf{b}_3 = 2\pi \frac{\mathbf{a}_1 \times \mathbf{a}_2}{\mathbf{a}_1 \cdot (\mathbf{a}_2 \times \mathbf{a}_3)}. \tag{2.32}$$

However, our main interests are in two dimensions for the description of the lattice of graphene and carbon nanotubes. Quite surprisingly, Eq. (2.32) requires some care to scale down to two dimensions primarily because the cross product is defined only in three (and some select higher) dimensions. To derive the reciprocal lattice vectors in two dimensions, let us consider a 2D plane defined by the vectors $\mathbf{a}_1$ and $\mathbf{a}_2$; the corresponding reciprocal vectors are $\mathbf{b}_1$ and $\mathbf{b}_2$.

---

[16] We have to train our attention to keep track of four lattice distinctions: the crystal lattice, the Bravais lattice, the direct lattice, and the reciprocal lattice. These terms can be mixed together to form stimulating brain teasers, because they sometimes mean the same thing. For example, the direct lattice is always a Bravais lattice, and the reciprocal lattice is itself always a Bravais lattice, but yet the reciprocal lattice is not the same as the direct lattice. Just like the nuances of any language, practice and familiarity are required, and afterwards it all makes sense.

[17] An educational derivation of the reciprocal lattice as the Fourier transform of the direct lattice can be found in several texts, including M. Dove, *Structure and Dynamics: An Atomic View of Materials* (Oxford University Press, 2003), Chapter 4.

Initially, we might be tempted to simply set $\mathbf{a}_3 = 0$, but the result is meaningless (the denominator of Eq. (2.32) vanishes). To resolve this dilemma, we need to rethink what two dimensions mean mathematically. Geometrically, all matter exists in a 3D space regardless of whether the matter is a skinny molecule or a gigantic mountain. When we speak of a 2D material or lattice, what is actually meant is that the lattice is restricted to a plane with *vanishing thickness*, where the vanishing thickness is an approximation for the relatively small thickness the lattice actually possesses (relative to its width and length). For the 2D plane of interest, the vanishing thickness approximation entails that the vectors $\mathbf{a}_1$ and $\mathbf{a}_2$ have zero components in the third dimension. Furthermore, the vector $\mathbf{a}_3$ exists only in the third dimension, that is, it is finite in the third dimension but vanishes in the other dimensions. Imposing these conditions on Eq. (2.32) allows it to be rewritten elegantly in two dimentions by utilizing vector operators:

$$\mathbf{b}_1 = 2\pi \frac{R_{90}(\mathbf{a}_2)}{\det(\mathbf{a}_1, \mathbf{a}_2)}, \qquad \mathbf{b}_2 = 2\pi \frac{R_{90}(-\mathbf{a}_1)}{\det(\mathbf{a}_1, \mathbf{a}_2)}, \qquad (2.33)$$

where R_{90} is an operator that rotates the vector clockwise by 90° (coordinate axis remains fixed) and det is the determinant, which geometrically is the area of the parallelogram formed by $\mathbf{a}_1$ and $\mathbf{a}_2$ and serves as a normalization factor. Since the rotation operator may not be familiar to some readers, let us look at a simple example. If vector $\mathbf{v} = v_1\hat{x} + v_2\hat{y}$, then $R_{90}(\pm\mathbf{v}) = \pm(v_2\hat{x} - v_1\hat{y})$. From the rotation operator, it is evident that the reciprocal lattice primitive vectors are either normal or parallel to the direct lattice primitive vectors, corresponding to $\mathbf{b}_i \cdot \mathbf{a}_j = 2\pi \delta_{ij}$, where δ_{ij} is the Kronecker delta function ($\delta_{ij} = 1$ if and only if $i = j$, and $\delta_{ij} = 0$ if $i \neq j$).

Some properties of the reciprocal lattice

(i) The reciprocal lattice is always a Bravais lattice, given that the direct lattice is always a Bravais lattice. However, it might be a different Bravais lattice compared with the original Bravais lattice in real space.

(ii) The reciprocal of the reciprocal lattice is the original direct lattice.

(iii) The direct lattice is measured in terms of a position vector $\mathbf{R}$ with units of length; the reciprocal lattice is measured in terms of the wavevector $\mathbf{K}$ with units of 1/length.

(iv) If s is the area of the direct lattice primitive cell, then $(2\pi)^2/s$ is the area of the reciprocal lattice primitive cell.

(v) The direct lattice exists in real space or position space; the reciprocal lattice exists in reciprocal space, which is sometimes called the Fourier space, momentum space, or simply k-space. This is because the crystal momentum is directly proportional to the wavevector.

(vi) The direct lattice can be viewed with a high-resolution electron microscope (as in Figure 2.7b), the reciprocal lattice is often characterized via X-ray diffraction patterns.

First Brillouin zone

The first Brillouin zone is a central concept in the theory of solids, particularly in the description of the band structure of electrons and other fundamental excitations[18] that exist within the solid. We previously discussed the first Brillouin zone in the context of the Kronig–Penney model. An equivalent but more basic definition is available based on the attributes of the reciprocal lattice: *the Wigner–Seitz primitive cell of the reciprocal lattice is the first Brillouin zone*. The standard convention is that the first Brillouin zone is the terminology exclusively used in the context of the reciprocal lattice, while the Wigner–Seitz terminology is used in the context of the direct lattice. Higher Brillouin zones also exist; however, those zones duplicate the information already present in the first Brillouin zone. As a result, casual usage of the term Brillouin zone often implies the first Brillouin zone in particular.

An example: the hexagonal lattice

Let us consider the 2D hexagonal lattice to demonstrate the construction of the reciprocal lattice and the first Brillouin zone. The hexagonal lattice in direct space and the associated Wigner–Seitz cell are shown in Figure 2.13a. A pair of symmetrical direct lattice primitive vectors from Eq. (2.30) are

$$\mathbf{a}_1 = \left(\frac{\sqrt{3}a}{2}, \frac{a}{2} \right), \qquad \mathbf{a}_2 = \left(\frac{\sqrt{3}a}{2}, -\frac{a}{2} \right).$$

The reciprocal lattice primitive vectors can be computed from Eq. (2.33):

$$\mathbf{b}_1 = \left(\frac{2\pi}{\sqrt{3}a}, \frac{2\pi}{a} \right), \qquad \mathbf{b}_2 = \left(\frac{2\pi}{\sqrt{3}a}, -\frac{2\pi}{a} \right). \qquad (2.34)$$

The resulting reciprocal lattice generated from sweeping $\mathbf{b}_1$ and $\mathbf{b}_2$ by integer multiples and the associated first Brillouin zone are shown in Figure 2.13b. The Wigner–Seitz cell and the first Brillouin zone are hexagons, displaying the hexagonal symmetry of the lattices. It is clearly visible from the figure that the reciprocal lattice and first Brillouin zone are rotated 90° with respect to the direct lattice.

---

[18] Besides electron excitations, there exist collective excitations such as lattice vibrations called phonons, electron density vibrations called plasmons, and spin vibrations called magnons. The Brillouin zone is useful in describing all these excitations.

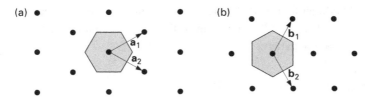

Fig. 2.13   (a) The hexagonal lattice and Wigner–Seitz cell (gray hexagon). (b) The corresponding reciprocal lattice, primitive vectors, and the first Brillouin zone (gray hexagon).

2.10   Summary

In this chapter we have explored at a very introductory level the quantum mechanical description of electrons in solids and the periodic structure found in crystalline solids. This introduction is more or less meant to serve as a refresher for those with prior quantum mechanics background. Nonetheless, the narrative is such that all the concepts and techniques developed are accessible to those with no prior course in quantum mechanics. The intuition developed and the understanding gained will be put to work in the development of the band structure of graphene and CNTs in Chapters 3 and 4 respectively.

2.11   Problem set

2.1. Electron scattering by a step potential barrier: a simple model to illustrate the ideas of reflection and transmission.

Consider the step potential barrier shown in Figure 2.14.

(a) Write the general solution for the electron wavefunction for each region (I, II). Note that the wavevector will be different in each region.

(b) What is the expression for k in each region in terms of the energy of that region?

(c) Now let us consider an electron wave in region I traveling towards the barrier with energy $E > U_0$. We want to determine the fraction of the electron that is reflected and transmitted by the barrier, a very important concept characteristic of all wave phenomena. What are the boundary conditions at $x = 0$?

(d) From the boundary conditions, determine the transmission and reflection probabilities, which are respectively defined as

$$T = \left| \frac{\psi_{II}^+}{\psi_{I}^+} \right|^2 \quad \text{and} \quad \Gamma = \left| \frac{\psi_{I}^-}{\psi_{I}^+} \right|^2, \quad \Gamma + T = 1,$$

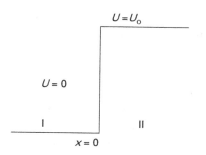

Fig. 2.14   Step potential barrier.

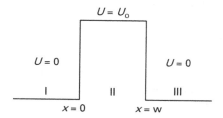

Fig. 2.15   Rectangular potential barrier.

where ψ_i^+ is a right-moving component of ψ_i and ψ_i^- is a left-moving component of ψ_i.

(e) Repeat parts (c) and (d) for an electron wave incident with energy $E < U_o$.

(f) For a barrier potential and width of $U_o = 0.3$ eV and $w = 1$ nm, plot the transmission probabilities for electron waves with $E > U_o$, and $E < U_o$.

(g) Assuming the electron is a classical particle like a ball and the potential barrier can be thought of as a wall, in this case sketch the transmission probability of the classical particle and superimpose it on the transmission probabilities from (f). What are the key differences, if any?

2.2. Electron scattering by a rectangular potential barrier: another simple model to illustrate the ideas of reflection and transmission.

Consider the potential barrier shown in Figure 2.15.

(a) Write the general solution for the electron wavefunction for each region (I, II, III). Note that the wavevector will be different in each region.

(b) What is the expression for k in each region in terms of the energy of that region?

(c) Now let us consider an electron in region I traveling towards the barrier with energy $E > U_o$. We want to determine the fraction of the electron that makes it through the barrier, a very important concept characteristic of all wave phenomena. What are the boundary conditions at $x = 0$ and $x = w$?

(d) From the boundary conditions, determine the transmission and reflection probabilities, which are respectively defined as:

$$T = \left|\frac{\psi_{\mathrm{III}}^{+}}{\psi_{\mathrm{I}}^{+}}\right|^2 \quad \text{and} \quad \Gamma = 1 - T,$$

where ψ_i^+ is a right-moving component of ψ_i.

(e) For a barrier potential and width of $U_0 = 0.3$ eV and $w = 1$ nm, plot the transmission probability.

(f) Assuming the electron is a classical particle like a ball and the potential barrier can be thought of as a wall, in this case sketch the transmission probability of the classical particle and compare it with your result from (e). What is the key difference, if any?

Bonus questions

(g) Newton's laws (classical mechanics) are often beaten up in quantum mechanics, but what is often not emphasized enough is that classical electromagnetism is still very useful. If you are familiar with Electromagnetism 101, you can solve a number of significant quantum mechanical problems almost by inspection by borrowing ideas from electromagnetic wave theory.

From part (e), determine the conditions that yield resonant transmission ($T = 1$ peaks). This is a familiar result from electromagnetic propagation.

(h) Solving the mathematics in quantum mechanics problems can often be tedious and "hard," but this is not the most important part of learning. More important is the physics behind the equations and results.

From part (g), what is the physical reason behind the resonant peaks? (*Hint*: There is a deep universal symmetry here: the reason for (h) is the same reason why there is resonance in all wave phenomena, such as electromagnetic, acoustic, seismic, and optical waves.)

2.3. A qualitative comparison of the bandgap of CNTs and graphene nanoribbons (GNRs).

You do not need to know what bandgap refers to in order to solve this problem. In brief, the bandgap is a fundamental property of all solids that informs us of whether the solid is an insulator, semiconductor, or metal; that is, how well it can conduct electricity. We are interested in comparing the bandgaps of CNTs and GNRs, at least qualitatively, to give us some basic insight as to which one has the lower or higher bandgap for the same dimensions. This is very important for electronic devices, as we will learn during the course of this book.

Consider an electron in a 1D empty solid of width w and infinite potential barriers at the edges. Qualitatively, this is the simplest model one can use for

CNTs or GNRs. It is certainly not accurate, but is good enough to provide important basic insights regarding bandgap dependency on the nanomaterial width.

(a) Write the general expression for the electron wavefunction.
   For GNRs we will employ the usual boundary conditions $\psi(0) = \psi(w) = 0$. For CNTs we will employ what is known as periodic boundary conditions applicable to conditions applicable when the beginning of the structure, touches the end of the structure such as a cylinder. The applicable boundary conditions are $\psi(0) = \psi(w) = e^{ikw}\psi(0)$.

(b) Determine the solution for the wavevector and the allowed energies in the GNR model.

(c) What is the lowest wavevector and energy of the electron? This energy can be thought of as directly related to the bandgap. Denote this energy as E_{g_min}.

(d) Repeat (b) for the CNT model.

(e) What is the lowest wavevector and energy of this electron in the CNT model? This energy can be thought of as directly related to the bandgap. Denote this energy as E_{c_min}.

(f) By what factor is E_{g_min} larger or smaller than E_{c_min}?

(g) So far, what we have been trying to determine is how the boundary conditions affect the bandgaps of nanotubes compared with nanoribbons. Another direct way to determine this is to look at experimental data or semi-empirical equations for the bandgap. For nanoribbons, consider the semi-empirical equation for the bandgap provided by Han et al.[19] For nanotubes, employ the relation $E_g(eV) = 0.9/d_t$ (nm). For a width of 20 nm, by what factor is the bandgap of nanotubes larger or smaller than nanoribbons? Compare with the factor determined in part (f).

2.4. The reciprocal lattice vectors.

(a) Prove that, given a set of primitive unit vectors $\mathbf{a}_1$, $\mathbf{a}_2$, and $\mathbf{a}_3$, the reciprocal lattice vectors defined in Eq. (2.32) satisfy the reciprocity condition of $e^{i\mathbf{K}\cdot\mathbf{R}} = 1$.

2.5. How many Bravais lattices are there in 1D space?

2.6. Crystal structure.

(a) Which of the 2D lattices in Figure 2.16 is not a Bravais lattice? A, B, C, or D?

(b) Determine its underlying Bravais lattice and draw two different primitive unit cells for that lattice. How many lattice points are there?

[19] M. Y. Han, B. Özyilmaz, Y. Zhang and P. Kim, Energy band-gap engineering of graphene nanoribbons. *Phys. Rev. Lett.*, **98** (2007) 206805.

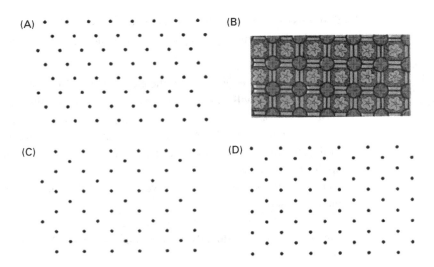

Fig. 2.16   The 2D lattices for Problem 2.6.

    (c) Construct a set of primitive reciprocal lattice vectors for that lattice.
    (d) Draw the reciprocal lattice.
    (e) Sketch the first Brillouin zone.

2.7. Bravais and reciprocal lattices.

    (a) Construct the reciprocal lattice for the five 2D Bravais lattices.
    (b) Construct the Brillouin zone for all five reciprocal lattices.
    (c) Which of the Bravais lattices yield hexagonal Brillouin zones?

3 Graphene

Dream your dreams and may they come true.

Felix Bloch (developed the theory describing
electrons in crystalline solids)

3.1 Introduction

The objective of this chapter is to describe the physical and electronic structure of graphene. Familiarity with concepts such as the crystal lattice and Schrödinger's quantum mechanical wave equation discussed in Chapter 2 will be useful. The electronic band structure of graphene is of primary importance because (i) it is the starting point for the understanding of graphene's solid-state properties and analysis of graphene devices and (ii) it is also the starting point for the understanding and derivation of the band structure of CNTs. We begin by broadly discussing carbon and then swiftly focus on graphene, including its crystal lattice and band structure. This chapter concludes on the contemporary topic of GNRs.

Carbon is a Group IV element that is very active in producing many molecular compounds and crystalline solids. Carbon has four valence electrons, which tend to interact with each other to produce the various types of carbon allotrope.[1] In elemental form, the four valence electrons occupy the 2s and 2p orbitals,[2] as illustrated in Figure 3.1a. When carbon atoms come together to form a crystal, one of the 2s electrons is excited to the $2p_z$ orbital from energy gained from neighboring nuclei, which has the net effect of lowering the overall energy of the system. Interactions or bonding subsequently follow between the 2s and 2p orbitals of neighboring carbon atoms. In chemistry, these interactions or mixing of atomic

---

[1] Allotropes are different structural modifications of an element in the same phase of matter, e.g. different solid forms.

[2] An orbital is essentially the electron wavefunction that informs us of the electron distribution around the nucleus. The elemental orbitals have specific distributions often graphed in standard chemistry texts. When elemental orbitals mix or hybridize, they form new orbitals with new electron distributions.

Table 3.1. Allotropes of graphene

Dimension	0D	1D	2D	3D
Allotrope	C_{60} buckyball	Carbon nanotubes	Graphene	Graphite
Structure	Spherical	Cylindrical	Planar	Stacked planar
Hybridization	sp^2	sp^2	sp^2	sp^2
Electronic properties	Semiconductor	Metal or semiconductor	Semi-metal	Metal

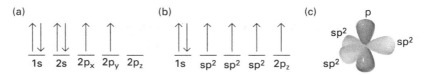

Fig. 3.1   The arrangement of electrons and their relative spin in (a) elemental carbon and (b) graphene. (b) The s and two of the p orbitals of the second shell interact covalently to form three sp^2 hybrid orbitals. (c) Illustration of the orbitals.

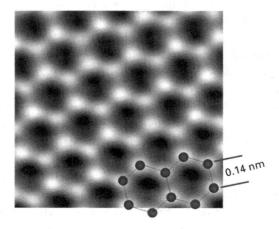

Fig. 3.2   Remarkable transmission electron aberration-corrected microscope (TEAM) image of graphene vividly showing the carbon atoms and bonds in the honeycomb structure (courtesy of Berkeley's TEAM05, 2009).

orbitals is commonly called hybridization, and the new orbitals that are formed are referred to as hybrid orbitals. The existence of multiple flavors of hybridization in carbon is what leads to the different allotropes shown in Table 3.1.

Graphene is a planar allotrope of carbon where all the carbon atoms form covalent bonds in a single plane. The planar honeycomb structure of graphene has been observed experimentally and is shown in Figure 3.2. Graphene can be considered the mother of three carbon allotropes. As illustrated in Figure 3.3, wrapping

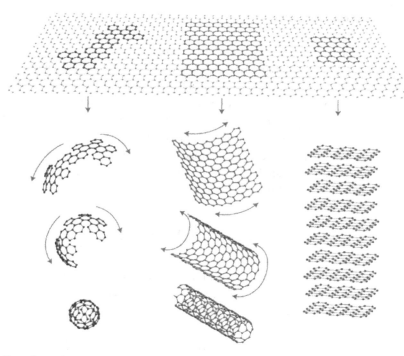

Fig. 3.3   Two-dimensional graphene can be considered the building block of several carbon allotropes in all dimensions, including zero-dimensional buckyballs, 1D nanotubes, and 3D graphite. Reprinted by permission from Macmillan Publishers Ltd (A. K. Geim and K. S. Novoselov, *Nat. Mater.*, **6**, (2007) 183–91, copyright (2007)).

graphene into a sphere produces buckyballs, folding into a cylinder produces nanotubes, and stacking several sheets of graphene leads to graphite. Furthermore, cutting graphene into a small ribbon results in nanoribbons, which are the subject of contemporary research. As a result, understanding the electronic properties of graphene is of central importance in explaining, for example, the electronic properties of carbon nanotubes and nanoribbons.

In graphene, the 2s orbital interacts with the $2p_x$ and $2p_y$ orbitals to form three sp^2 hybrid orbitals with the electron arrangement shown in Figure 3.1b. The sp^2 interactions result in three bonds called σ-bonds, which are the strongest type of covalent bond. The σ-bonds have the electrons localized along the plane connecting carbon atoms and are responsible for the great strength and mechanical properties of graphene and CNTs. The $2p_z$ electrons forms covalent bonds called π-bonds, where the electron cloud is distributed normal to the plane connecting carbon atoms. The $2p_z$ electrons are weakly bound to the nuclei and, hence, are relatively delocalized. These delocalized electrons are the ones responsible for the electronic properties of graphene and CNTs and as such will occupy much of our attention.

3.2   The direct lattice

Graphene has a honeycomb lattice shown in Figure 3.4 using a *ball-and-stick* model. The balls represent carbon atoms and the sticks symbolize the σ-bonds between atoms. The carbon–carbon bond length is approximately $a_{C-C} \approx 1.42$ Å. The honeycomb lattice can be characterized as a Bravais lattice with a basis of two atoms, indicated as A and B in Figure 3.4, and these contribute a total of two π electrons per unit cell to the electronic properties of graphene. The underlying Bravais lattice is a hexagonal lattice and the primitive unit cell can be considered an equilateral parallelogram with side $a = \sqrt{3}a_{C-C} = 2.46$ Å. The primitive unit vectors as defined in Figure 3.4 are

$$\mathbf{a}_1 = \left(\frac{\sqrt{3}a}{2}, \frac{a}{2}\right), \quad \mathbf{a}_2 = \left(\frac{\sqrt{3}a}{2}, \frac{a}{-2}\right), \tag{3.1}$$

with $|\mathbf{a}_1| = |\mathbf{a}_2| = a$. Each carbon atom is bonded to its three nearest neighbors and the vectors describing the separation between a type A atom and the nearest neighbor type B atoms as shown in Figure 3.4 are

$$\mathbf{R}_1 = \left(\frac{a}{\sqrt{3}}, 0\right), \mathbf{R}_2 = -\mathbf{a}_2 + \mathbf{R}_1 = \left(-\frac{a}{2\sqrt{3}}, +\frac{a}{2}\right),$$
$$\mathbf{R}_3 = -\mathbf{a}_1 + \mathbf{R}_1 = \left(-\frac{a}{2\sqrt{3}}, -\frac{a}{2}\right), \tag{3.2}$$

with $|\mathbf{R}_1| = |\mathbf{R}_2| = |\mathbf{R}_3| = a_{C-C}$.

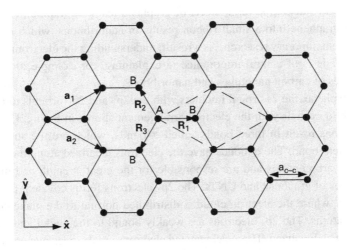

Fig. 3.4   The honeycomb lattice of graphene. The primitive unit cell is the equilateral parallelogram (dashed lines) with a basis of two atoms denoted as A and B.

3.3 The reciprocal lattice

The reciprocal lattice of graphene shown in Figure 3.5 is also a hexagonal lattice, but rotated 90° with respect to the direct lattice. The reciprocal lattice vectors are (from Eq. (2.34))

$$\mathbf{b}_1 = \left(\frac{2\pi}{\sqrt{3}a}, \frac{2\pi}{a} \right), \quad \mathbf{b}_2 = \left(\frac{2\pi}{\sqrt{3}a}, -\frac{2\pi}{a} \right), \tag{3.3}$$

with $|\mathbf{b}_1| = |\mathbf{b}_2| = 4\pi/\sqrt{3}a$. The Brillouin zone, which is a central idea in describing the electronic bands of solids, is illustrated as the shaded hexagon in Figure 3.5 with sides of length $b_{BZ} = |b_1|/\sqrt{3} = 4\pi/3a$ and area equal to $8\pi^2/\sqrt{3}a^2$. There are three key locations of high symmetry in the Brillouin zone which are useful to memorize in discussing the dispersion of graphene. In Figure 3.5, these locations are identified by convention as the Γ-point, the M-point, and the K-point.[3] The Γ-point is at the center of the Brillouin zone, and the vectors describing the location of the other points with respect to the zone center are

$$\Gamma M = \left(\frac{2\pi}{\sqrt{3}a}, 0 \right), \quad \Gamma K = \left(\frac{2\pi}{\sqrt{3}a}, \frac{2\pi}{3a} \right), \tag{3.4}$$

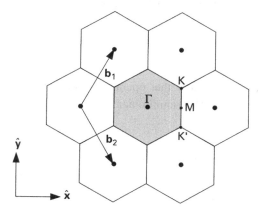

Fig. 3.5 The reciprocal lattice of graphene. The first Brillouin zone is the shaded hexagon with the high symmetry points labeled as Γ, M, and K located at the center, midpoint of the side, and corner of the hexagon respectively. The K'-point, which is also an hexagonal corner (adjacent to the K-point), is essentially equivalent to the K-point for most purposes.[4]

[3] The naming of the high-symmetry points for every Bravais lattice has its origins in group theory; for reference, see M. S. Dresselhaus, G. Dresselhaus and A. Jorio, *Group Theory: Application to the Physics of Condensed Matter* (Springer, 2008).

[4] Sometimes a distinction is made between the K-point and K'-point, particularly in the discussion of intervalley or interband electron scattering by lattice vibrations (more about this in Chapter 7). For our current studies, we will simply call out the K-points to refer to all the corners of the hexagonal Brillouin zone unless explicitly stated otherwise.

with $|\Gamma M| = 2\pi/\sqrt{3}a$, $|\Gamma \mathbf{K}| = 4\pi/3a$, and $|M\mathbf{K}| = 2\pi/3a$. There are six K-points[4] and six M-points within the Brillouin zone.

The unique solutions for the energy bands of crystalline solids are found within the Brillouin zone and sometimes the dispersion is graphed along the high-symmetry directions as a matter of practical convenience. Furthermore, we shall sometimes use the terminology k-space to refer to the reciprocal lattice and the vector that locates any point within the Brillouin zone is the wavevector $\mathbf{k}$. That is, every allowed point (also synonymous with the term *allowed state* when referring to the dispersion) within the Brillouin zone can be reached by $\mathbf{k}$.

3.4   Electronic band structure

The band structure of graphene is shown in Figure 3.6. This band structure was computed numerically from first principles and shows many energy branches resulting from all the π and σ electrons that form the outermost electrons of carbon. Our purpose here is to develop an analytical description of the band structure that would nurture our intuition for the device physics and lead to simple algebraic expressions for the relevant device and material parameters. Invariably, developing an analytical expression for the band structure of a solid generally requires solving the time-independent Schrödinger's equation in 3D space:

$$H\Psi(k, r) = E(k)\Psi(k, r), \tag{3.5}$$

where H is the Hamiltonian operator that operates on the wavefunction Ψ to produce the allowed energies E. The Hamiltonian for an independent electron in a periodic solid is given by

$$H = \frac{\hbar^2}{2m}\nabla^2 + \Sigma_i^N U(\mathbf{r} - \mathbf{R}_i), \tag{3.6}$$

where the former term in the Hamiltonian is the kinetic energy operator and the latter term is the potential energy operator. $\mathbf{R}_i$ is the ith Bravais lattice vector, N is the number of primitive unit cells, and $U(\mathbf{r}-\mathbf{R}_i)$ is the potential energy contribution from the atom centered in the ith primitive unit cell. The potential energy is the sum of the single-atom potentials and, hence, is periodic and Coulombic in nature ($1/r$ dependence). Substituting the Hamiltonian into the Schrödinger equation results in a second-order partial differential equation. In practice, constructing a wavefunction that satisfies the partial differential equation is non-trivial, requiring deep thought, and may consist of iterating over several trial wavefunctions until a suitable wavefunction is determined. In any case, acceptable wavefunctions in a crystalline solid must satisfy Bloch's theorem, which proves that valid traveling

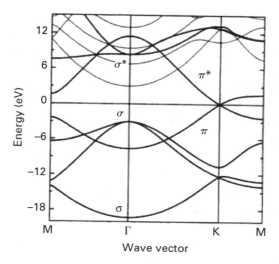

Fig. 3.6 The *ab-initio* band structure of graphene, including the σ and π bands. The Fermi energy is set to 0 eV. Adapted with permission from M. Machón *et al.*, *Phys. Rev. B*, **66**, (2002) 155410. Copyright (2002) by the American Physical Society.

waves in a lattice have the property

$$\Psi(\mathbf{r} + \mathbf{R}) = e^{i\mathbf{k}\cdot\mathbf{R}}\Psi(\mathbf{r}), \tag{3.7}$$

where $\mathbf{R}$ is a Bravais lattice vector. In addition, periodic boundary conditions are imposed on the wavefunctions to determine the allowed values of the wavevector that leads to running waves:

$$\Psi(\mathbf{r}) = \Psi(\mathbf{r} + \mathbf{S}) = e^{i\mathbf{k}\cdot\mathbf{S}}\Psi(\mathbf{r}), \Rightarrow \therefore e^{i\mathbf{k}\cdot\mathbf{S}} = 1, \tag{3.8}$$

where $\mathbf{S}$ is the size vector whose lengths in all the coordinates of space are the spatial dimensions of the lattice.

In general there are two limiting techniques for obtaining a satisfactory wavefunction and associated band structure.[5] In one limiting case, called the nearly-free electron model, the outermost valence electrons are essentially considered free except for a weak Coulomb attraction to their respective nucleus. As a result, the

---

[5] Both techniques (nearly-free electron and tight-binding models) were developed by Felix Bloch during his Ph.D. research in the late 1920s. Bloch was very fluent in Fourier transforms and his application of Fourier analysis to electrons in solids gave him original insights into their properties. Bloch later went on to become the first Nobel laureate at Stanford University in 1952.

full periodic potential is replaced by a weak perturbing potential and Schrödinger's equation is solved by employing standard perturbation techniques in quantum mechanics. This model yields solutions in terms of modulated plane waves ($\Psi \sim u(r)e^{ik \cdot r}$), and the associated energy bands often have a parabolic structure. This model has been shown to be useful in describing the band structure of some metals.

The other limiting technique, called the tight-binding model, inherently assumes that the outermost electrons are to a large extent localized (i.e. tightly bound) to their respective atomic cores and, hence, described by their atomic orbitals with discrete energy levels. However, because the atoms are not isolated but exist in an ordered solid, the orbitals of identical electrons in neighboring atoms in a solid with N unit cells will overlap, with the major consequence that the N discrete energy levels will inevitably broaden into quasi-continuous energy bands with N states/band owing to Pauli's exclusion principle. The overlap of wavefunctions by and large renders inaccurate the use of atomic orbitals in describing electrons in a solid. Nonetheless, for the special case of a very small overlap, one might still be able to use the tight-binding model to obtain an approximate analytical band structure that we hope will be in good agreement with experimental measurements or more sophisticated numerical *ab-initio* band structure computations.

We have to choose between these two models in order to develop an analytical electronic band structure. For the particular case of graphene, a variety of arguments can be proposed (mostly based on experience, because band structure calculation is as much an art as it is science) to support the choice of a particular model. Fortunately, we know from chemistry that graphene can be considered a large carbon molecule and, as such, a first guess might be to employ standard quantum chemistry techniques such as linear combination of atomic orbitals (which we have been calling tight binding) for deriving molecular band structures. Furthermore, visual inspection of the *ab-initio* computations (Figure 3.6), particularly around the Fermi energy (E_F, the energy at 0 eV),[6] shows a linear dispersion, suggesting that perhaps a nearly-free electron model might not be our first choice, since that would require a large number of plane waves. If the dispersion had been parabolic-like, a nearly-free electron model might arguably be a more attractive initial choice. Accordingly, we chose the tight-binding model. How well the tight-binding model agrees with *ab-initio* computations or experimental data is the ultimate judge of whether the chosen model is indeed useful. We will now proceed to dive into the detailed mathematics and derive the tight-binding band structure of graphene.

---

[6] The most *mobile* electrons are at the Fermi energy. We will use $E_F = 0$ eV casually for now until later, at the end of this section, when a formal definition is presented to identify its location in the band structure of graphene.

3.5   Tight-binding energy dispersion

Rather than discussing very broadly the tight-binding formalism, we will focus on the specific problem of calculating the band structure of graphene. The primary challenge in the tight-binding model is to *construct* a suitable wavefunction that satisfies Bloch's theorem while retaining the atomic character. From this wavefunction, the subsequent calculation of the energy bands is fairly straightforward, though it can be mathematically tedious. Fortunately, a general tight-binding *ansatz*[7] for the wavefunction has been constructed previously in terms of linear combinations of atomic orbitals originally proposed by Bloch in 1928. For graphene, which has a basis of two, the tight-binding wavefunction is a weighted sum of the two sub-lattice Bloch functions:

$$\Psi(\mathbf{k}, \mathbf{r}) = C_A \Phi_A(\mathbf{k}, \mathbf{r}) + C_B \Phi_B(\mathbf{k}, \mathbf{r}), \tag{3.9}$$

where the subscripts A and B denote the two different atoms in the graphene unit cell (see Figure 3.4). The weights (C_A, C_B) are, in general, functions of $\mathbf{k}$ but independent of $\mathbf{r}$. For a crystal with a basis of m, the sum will include m sub-lattice terms. The *ansatz* expresses the Bloch functions as a linear combination of the atomic orbitals or wavefunctions which are assumed to be known:

$$\Phi_A(\mathbf{k}, \mathbf{r}) = \frac{1}{\sqrt{N}} \Sigma_j^N e^{i\mathbf{k} \cdot \mathbf{R}_{A_j}} \phi(\mathbf{r} - \mathbf{R}_{A_j}) \tag{3.10}$$

$$\Phi_B(\mathbf{k}, \mathbf{r}) = \frac{1}{\sqrt{N}} \Sigma_j^N e^{i\mathbf{k} \cdot \mathbf{R}_{B_j}} \phi(\mathbf{r} - \mathbf{R}_{B_j}), \tag{3.11}$$

where N is the number of unit cells in the lattice and $\mathbf{R}_A (\mathbf{R}_B)$ are the Bravais lattice vectors identifying the locations of all type A (B) atoms in the graphene lattice. The atomic orbitals ϕ belong to a class of functions known as Wannier functions, which are orthonormal functions that are sufficiently localized such that, at distances increasingly removed from the center point $\mathbf{R}_j$, the functions decay to zero very rapidly. The sum is over all the lattice vectors, and $1/\sqrt{N}$ serves as normalization constant for the Bloch functions in the strict limit when the Wannier function in cell j has zero overlap with neighboring Wannier functions.[8]

[7] *Ansatz* is a German word for an educated guess whose validity is based on the accuracy of its predictions. Observe that the wavefunction is not computed directly from solving Schrödinger's equation, but simply postulated as meeting basic requirements such as Bloch's theorem. Use of an *ansatz* is a common technique in quantum physics to describe constructions of solutions to non-trivial differential equations.

[8] The Bloch functions are not exact wavefunctions because they are not normalized when we include some finite overlap. However, they are the best tight-binding *ansatz* we have that is not overly complicated and is still suitable for analysis with useful accuracy.

The Bloch functions must satisfy Bloch's theorem stated in Eq. (3.7):

$$\Phi_A(\mathbf{r} + \mathbf{R}_{A\ell}) = \frac{1}{\sqrt{N}} \Sigma_j^N e^{i\mathbf{k}\cdot\mathbf{R}_{Aj}} \phi(\mathbf{r} + \mathbf{R}_{A\ell} - \mathbf{R}_{Aj})$$

$$\Phi_A(\mathbf{r} + \mathbf{R}_{A\ell}) = \frac{e^{i\mathbf{k}\cdot\mathbf{R}_{A\ell}}}{\sqrt{N}} \Sigma_j^N e^{i\mathbf{k}\cdot(\mathbf{R}_{Aj} - \mathbf{R}_{A\ell})} \phi(\mathbf{r} - (\mathbf{R}_{Aj} - \mathbf{R}_{A\ell})). \qquad (3.12)$$

The difference between two Bravais lattice vectors is another Bravais lattice vector; therefore:

$$\Phi_A(\mathbf{r} + \mathbf{R}_{A\ell}) = \frac{e^{i\mathbf{k}\cdot\mathbf{R}_{A\ell}}}{\sqrt{N}} \Sigma_m^N e^{i\mathbf{k}\cdot\mathbf{R}_{Am}} \phi(\mathbf{r} - \mathbf{R}_{Am}) = e^{i\mathbf{k}\cdot\mathbf{R}_{A\ell}} \Phi_A(\mathbf{r}). \qquad (3.13)$$

Additionally, the Bloch functions must satisfy the periodic boundary conditions stated in Eq. (3.8). Let us express the reciprocal lattice variable $\mathbf{k}$ in terms of its coordinate components, $\mathbf{k} = k_x\mathbf{x} + k_y\mathbf{y}$, and let the size vector of the graphene Bravais lattice be $\mathbf{S} = a\mathrm{No}\mathbf{x} + a\mathrm{No}\mathbf{y}$, where $\mathrm{No} = \sqrt{N}$. Imposing the periodic boundary conditions

$$e^{i\mathbf{k}\cdot\mathbf{S}} = \cos[ak_x\mathrm{No} + ak_y\mathrm{No}] + i \sin[ak_x\mathrm{No} + ak_y\mathrm{No}] = 1, \qquad (3.14)$$

which is only satisfied when

$$k_x = \frac{2\pi p}{a\mathrm{No}}, \quad k_y = \frac{2\pi p}{a\mathrm{No}}, \quad p = 0, 1, 2, \ldots, \mathrm{No} - 1, \qquad (3.15)$$

where the allowed set of wavevectors that yield unique solutions for the energy dispersion are limited to the first Brillouin zone. The maximum number of k-states in the Brillouin zone is $\mathrm{No}^2 = N$, which can hold $2N$ electrons, where the factor of 2 is due to spin degeneracy.

With the formalities out of the way, we can now proceed to solve for the energy bands of graphene. Inserting Eq. (3.9) into the Schrödinger equation we obtain

$$C_A H \Phi_A(\mathbf{k}, \mathbf{r}) + C_B H \Phi_B(\mathbf{k}, \mathbf{r}) = E(\mathbf{k})C_A \Phi_A(\mathbf{k}, \mathbf{r}) + E(\mathbf{k})C_B \Phi_B(\mathbf{k}, \mathbf{r}). \qquad (3.16)$$

Multiplying by the complex conjugate of Φ_A, and separately by the complex conjugate of Φ_B, generates two separate equations:[9]

$$C_A \Phi_A^* H \Phi_A + C_B \Phi_A^* H \Phi_B = E C_A \Phi_A^* \Phi_A + E C_B \Phi_A^* \Phi_B$$

$$C_A \Phi_B^* H \Phi_A + C_B \Phi_B^* H \Phi_B = E C_A \Phi_B^* \Phi_A + E C_B \Phi_B^* \Phi_B \qquad (3.17)$$

---

[9] As a refresher, mathematics is very precisely formulated in quantum mechanics, including the order of multiplication, because the operations may not be commutative.

where we have dropped the dependence on $\mathbf{k}$ and $\mathbf{r}$ for convenience. Integrating both equations over the entire space occupied by the lattice (denoted by Ω) produces[10]

$$C_A \int_\Omega \Phi_A^* H \Phi_A d\mathbf{r} + C_B \int_\Omega \Phi_A^* H \Phi_B d\mathbf{r} = EC_A \int_\Omega \Phi_A^* \Phi_A d\mathbf{r} + EC_B \int_\Omega \Phi_A^* \Phi_B d\mathbf{r}$$

$$C_A \int_\Omega \Phi_B^* H \Phi_A d\mathbf{r} + C_B \int_\Omega \Phi_B^* H \Phi_B d\mathbf{r} = EC_A \int_\Omega \Phi_B^* \Phi_A d\mathbf{r} + EC_B \int_\Omega \Phi_B^* \Phi_B d\mathbf{r}.$$

$$(3.18)$$

It is customary to employ the following symbolic definitions to make the equations more manageable:

$$H_{ij} = \int_\Omega \Phi_i^* H \Phi_j d\mathbf{r}, \quad S_{ij} = \int_\Omega \Phi_i^* \Phi_j d\mathbf{r}, \tag{3.19}$$

where H_{ij} are the matrix elements of the Hamiltonian or transfer integral and have the units of energy. S_{ij} are the overlap matrix elements between Bloch functions and are unitless. We can simplify the matrix elements by observing that, since the two atoms in the unit cell are identical, therefore, the overlap between all type-A atoms must be the same as the overlap between all type-B atoms; that is, $S_{AA} = S_{BB}$ and $H_{AA} = H_{BB}$. In addition, H_{ij} and S_{ij} correspond to physical observables and hence are Hermitian, which leads to the condition $H_{BA} = H_{AB}^*$ and $S_{BA} = S_{AB}^*$. Then:

$$C_A(H_{AA} - ES_{AA}) = C_B(ES_{AB} - H_{AB}) \tag{3.20}$$

$$C_A(H_{AB}^* - ES_{AB}^*) = C_B(ES_{AA} - H_{AA}). \tag{3.21}$$

Our primary interest here is to obtain an expression for the energy bands. This system of two linear equations can be solved easily by the method of substitution. Solving for C_B in Eq. (3.21) and substituting into Eq. (3.20) yields a quadratic equation, which is readily solved for the energy:

$$E(\mathbf{k})^\pm = \frac{E_0(\mathbf{k}) \pm \sqrt{E_0(\mathbf{k})^2 - 4(S_{AA}(\mathbf{k})^2 - |S_{AB}(\mathbf{k})|^2)(H_{AA}(\mathbf{k})^2 - |H_{AB}(\mathbf{k})|^2)}}{2(S_{AA}(\mathbf{k})^2 - |S_{AB}(\mathbf{k})|^2)},$$

$$(3.22)$$

---

[10] It is worthwhile remembering that the integration of pairs of a conjugated wavefunction is what has real probabilistic meaning.

with

$$E_o(\mathbf{k}) = (2H_{AA}(\mathbf{k})S_{AA}(\mathbf{k}) - S_{AB}(\mathbf{k})H^*_{AB}(\mathbf{k}) - H_{AB}(\mathbf{k})S^*_{AB}(\mathbf{k})). \qquad (3.23)$$

The positive and negative energy branches in Eq. (3.22) are called the conduction (π^*) and valence (π) bands respectively,[11] and will be defined to correspond to positive and negative energy values in a while. Up till now we have not fully explored *tight binding* in the mathematical development of the energy bands. To make further progress in Eq. (3.22), we will now make explicit use of the assumptions belonging to the tight-binding formalism. Additionally, since our goal is to develop an analytical equation for the band structure that will resemble the *ab-initio* computation in Figure 3.6 (*particularly around the Fermi energy*), insightful visual observations of Figure 3.6 will be brought to bear on the problem at hand in order to arrive at a final solution for the energy bands of graphene. The assumptions and related mathematical consequences are:

1. *Nearest neighbor tight-binding (NNTB) model*: The wavefunction of an electron in any primitive unit cell *only* overlaps with the wavefunctions of its *nearest neighbors*. This is easily understood by means of Figure 3.4. The nearest neighbors of a type-A atom in the graphene lattice are three equivalent type-B atoms. NNTB stipulates that the p_z wavefunction of a type-A atom overlaps with the p_z wavefunctions of its three nearest neighbors, and zero overlap with wavefunctions from farther atoms. Mathematically, this simplifies Eq. (3.22) considerably, as the Hamiltonian matrix element reduces to

$$H_{AA}(\mathbf{k}) = \int_{\Omega} \Phi^*_A H \Phi_A dr = \frac{1}{N} \sum_{j}^{N} \sum_{l}^{N} e^{-i\mathbf{k}\cdot\mathbf{R}_{A_j}} e^{i\mathbf{k}\cdot\mathbf{R}_{A_l}}$$

$$\times \int_{\Omega} \phi^*(\mathbf{r} - \mathbf{R}_{A_j}) H \phi(\mathbf{r} - \mathbf{R}_{A_l}) dr \qquad (3.24)$$

$$H_{AA} = \frac{1}{N} \sum_{j}^{N} \sum_{l}^{N} e^{i\mathbf{k}\cdot(\mathbf{R}_{A_l} - \mathbf{R}_{A_j})} E_{2p}\delta_{jl} = E_{2p}, \qquad (3.25)$$

where δ_{jl} is the Kronecker delta function. The constant term E_{2p} is nominally close to the energy of the 2p orbital in isolated carbon, but not exactly the same because the Hamiltonian of the lattice (Eq. (3.6)) has a periodic potential, in contrast to the single Columbic potential of the isolated atom. Similarly, the

---

[11] The π and π^* are the preferred terminology in chemistry, used to describe bonding and anti-bonding interactions (or energies) respectively. The anti-bonding interactions are higher in energy than the bonding interactions.

overlap matrix reduces to

$$S_{AA}(\mathbf{k}) = \int_{\Omega} \Phi_A^* \Phi_A dr = \frac{1}{N} \sum_j^N \sum_l^N e^{-i\mathbf{k}\cdot\mathbf{R}_{A_j}} e^{i\mathbf{k}\cdot\mathbf{R}_{A_l}}$$

$$\times \int_{\Omega} \phi^*(\mathbf{r} - \mathbf{R}_{A_j})\phi(\mathbf{r} - \mathbf{R}_{A_l}) dr \qquad (3.26)$$

$$S_{AA} = \frac{1}{N} \sum_j^N \sum_l^N e^{i\mathbf{k}\cdot(\mathbf{R}_{A_l} - \mathbf{R}_{A_j})} \delta_{jl} = 1, \qquad (3.27)$$

where we have taken advantage of the normalized feature of Wannier functions, $\int \phi^*(\mathbf{r} - \mathbf{R}_j)\phi(\mathbf{r} - \mathbf{R}_j)dr = 1$.

2. *Electron–hole symmetry*: A close observation of the *ab-initio* dispersion in the neighborhood of the Fermi energy ($E = 0$ at the K-point in Figure 3.6) reveals that the π and π^* branches have similar structure, at least for energies close to the E_F. Within this restricted range, the energy branches are approximately mirror images of each other. Since electrons are the mobile charges in the π^* band and holes are the mobile charges in the π band, we call this approximation the electron–hole symmetry.[12] Obviously, from Figure 3.6 this symmetry does not hold over a large range of energies. Nonetheless, this approximation is very useful because much of the electron dynamics in practical devices occur over a relatively small range of energies close to the Fermi energy. Mathematically, electron–hole symmetry forces $S_{AB}(\mathbf{k}) = 0$. This is not immediately obvious, but can be seen (after some deep thought) by observing that the only part of Eq. (3.22) that possesses symmetry about some number is the plus/minus square root argument, i.e. $-\sqrt{f(\mathbf{k})}$ is a mirror image of $+\sqrt{f(\mathbf{k})}$. In order for Eq. (3.22) to retain this symmetry, $S_{AB}(\mathbf{k})$ must vanish to zero. Therefore:

$$E(\mathbf{k})^{\pm} = E_{2p} \pm \sqrt{H_{AB}(\mathbf{k})H_{AB}^*(\mathbf{k})}, \qquad (3.28)$$

which is the energy dispersion originally proposed by Wallace in 1947.[13] The electron–hole symmetry argument is somewhat subtle yet powerful in simplifying Eq. (3.22) without compromising on accuracy, particularly at energies close to E_F.

We can go one step further to simplify Eq. (3.28) without any loss of accuracy by setting $E_{2p} = 0$. This is because energy is naturally defined to within an arbitrary reference potential. For the case of graphene, the reference potential is the Fermi energy (a parameter independent of $\mathbf{k}$) and is customarily set to 0 eV. Since the

---

[12] A key physical outcome of this symmetry is that electrons and holes will have identical equilibrium properties, such as density of states, group velocity, and carrier density.
[13] P. R. Wallace, The band theory of graphite. *Phys. Rev.*, **71** (1947) 622–34.

only parameter independent of $\mathbf{k}$ in Eq. (3.28) is E_{2p}, it is convenient to employ it as the reference and hence $E_{2p} = E_F = 0$ eV. Finally, we arrive at

$$E(\mathbf{k})^{\pm} = \pm\sqrt{H_{AB}(\mathbf{k})H_{AB}^*(\mathbf{k})}. \tag{3.29}$$

The Hamiltonian matrix element $H_{AB}(\mathbf{k})$ can be calculated in a straightforward manner, though is somewhat algebraically tedious:

$$H_{AB}(\mathbf{k}) = \int_{\Omega} \Phi_A^* H \Phi_B dr = \frac{1}{N} \sum_{j}^{N} \sum_{l}^{N} e^{-i\mathbf{k}\cdot(\mathbf{R}_{Aj}-\mathbf{R}_{Bl})}$$

$$\times \int_{\Omega} \phi^*(\mathbf{r} - \mathbf{R}_{Aj}) H \phi(\mathbf{r} - \mathbf{R}_{Bl}) dr \tag{3.30}$$

As a result of the NNTB stipulations, each type-A atom will only overlap with its three nearest neighbors, which are type B. It is convenient to define the nearest neighbor distances between a type-A atom and its three type-B atoms. Let these nearest neighbor distances be $\mathbf{R}_1 = \mathbf{R}_{Aj} - \mathbf{R}_{Bj}$, $\mathbf{R}_2 = \mathbf{R}_{Aj} - \mathbf{R}_{Bj+1}$, $\mathbf{R}_3 = \mathbf{R}_{Aj} - \mathbf{R}_{Bj-1}$ (as denoted in Figure 3.4), where $j, j+1$, and $j-1$ are the indices of the primitive unit cells where the three type-B nearest neighbor atoms are located with respect to atom A in cell j.

$$H_{AB}(\mathbf{k}) = \frac{1}{N} \sum_{j}^{N} \sum_{m=1}^{3} e^{-i\mathbf{k}\cdot\mathbf{R}_m} E_m \tag{3.31}$$

where E_m is the finite value of the integration of the nearest neighbor Wannier functions. By necessity $E_1 = E_2 = E_3$, because the integrals are radially dependent and the nearest neighbor distances are radially symmetric. For convenience we can simply set $E_m = \gamma$, which can be viewed as a fitting parameter of positive value.[14] γ is often called by many names, including the nearest neighbor overlap energy, the hopping or transfer energy, or the carbon–carbon interaction energy. The Hamiltonian matrix element reduces to a sum of three terms:

$$H_{AB}(\mathbf{k}) = \gamma(e^{-i\mathbf{k}\cdot\mathbf{R}_1} + e^{-i\mathbf{k}\cdot\mathbf{R}_2} + e^{-i\mathbf{k}\cdot\mathbf{R}_3}) \tag{3.32}$$

and

$$H_{AB}(\mathbf{k})H_{AB}^*(\mathbf{k}) = \gamma^2(e^{-i\mathbf{k}\cdot\mathbf{R}_1} + e^{-i\mathbf{k}\cdot\mathbf{R}_2} + e^{-i\mathbf{k}\cdot\mathbf{R}_3})(e^{i\mathbf{k}\cdot\mathbf{R}_1} + e^{i\mathbf{k}\cdot\mathbf{R}_2} + e^{i\mathbf{k}\cdot\mathbf{R}_3}), \tag{3.33}$$

---

[14] Strictly speaking γ is a negative value owing to the sign convention for attractive potentials, and some authors define it accordingly. Here, it is defined positive for convenience. Ultimately, the polarity does not matter, provided the analysis is consistent.

$$H_{AB}(\mathbf{k})H_{AB}^*(\mathbf{k}) = \gamma^2[3 + e^{i\mathbf{k}\cdot(\mathbf{R}_1-\mathbf{R}_2)} + e^{-i\mathbf{k}\cdot(\mathbf{R}_1-\mathbf{R}_2)} + e^{i\mathbf{k}\cdot(\mathbf{R}_1-\mathbf{R}_3)}$$
$$+ e^{-i\mathbf{k}\cdot(\mathbf{R}_1-\mathbf{R}_3)} + e^{i\mathbf{k}\cdot(\mathbf{R}_2-\mathbf{R}_3)} + e^{-i\mathbf{k}\cdot(\mathbf{R}_2-\mathbf{R}_3)}], \tag{3.34}$$

$$H_{AB}(\mathbf{k})H_{AB}^*(\mathbf{k}) = \gamma^2\{3 + 2\cos[\mathbf{k}\cdot(\mathbf{R}_1-\mathbf{R}_2)] + 2\cos[\mathbf{k}\cdot(\mathbf{R}_1-\mathbf{R}_3)]$$
$$+ 2\cos[\mathbf{k}\cdot(\mathbf{R}_2-\mathbf{R}_3)]\}, \tag{3.35}$$

where Euler's formula, $e^{ix} + e^{-ix} = 2\cos[x]$, has been used to simplify the exponentials. The nearest neighbor distances are given by Eq. (3.2); thus:

$$H_{AB}(\mathbf{k})H_{AB}^*(\mathbf{k}) = \gamma^2\left(1 + 4\cos\frac{\sqrt{3}a}{2}k_x\cos\frac{a}{2}k_y + 4\cos^2\frac{a}{2}k_y\right). \tag{3.36}$$

And hence our central mission in deriving analytically the π bands of graphene can now be accomplished by inserting Eq. (3.36) into Eq. (3.29):

$$E(\mathbf{k})^\pm = \pm\gamma\sqrt{1 + 4\cos\frac{\sqrt{3}a}{2}k_x\cos\frac{a}{2}k_y + 4\cos^2\frac{a}{2}k_y}. \tag{3.37}$$

This formula is the widely used NNTB approximation from which CNT band structure is derived. The precise value of γ is difficult to determine analytically; as such, it is often used as a fitting parameter to match *ab-initio* computations or experimental data. Commonly used values for γ range from about 2.7 eV to 3.3 eV. For routine calculations, the extracted value $\gamma \approx 3.1$ eV from experimental measurements of the Fermi (or group) velocity ($v_F \approx 10^6$ m s^{-1}) in graphene is adequate.[15] It is particularly of interest to identify the highest energy state within the valence band and the lowest energy state within the conduction band. These states occur at the K-point corresponding to $E = 0$ eV.

Comparison of the NNTB dispersion, Eq. (3.37), with *ab-initio* computations for the π bands shows good agreement (Figure 3.7) with the strongest agreement expectedly at low energies.[16] A greater agreement at higher energies can be achieved by relaxing the electron–hole symmetry approximation, which leads to a non-zero $S_{AB}(\mathbf{k})$. For this case, it can be shown that the energy dispersion is

$$E(\mathbf{k})^\pm = \frac{\pm\gamma\sqrt{1 + 4\cos\frac{\sqrt{3}a}{2}k_x\cos\frac{a}{2}k_y + 4\cos^2\frac{a}{2}k_y}}{1 \mp s_0\sqrt{1 + 4\cos\frac{\sqrt{3}a}{2}k_x\cos\frac{a}{2}k_y + 4\cos^2\frac{a}{2}k_y}} \tag{3.38}$$

[15] Theoretically, v_F is proportional to γ. See R. S. Deacon, K.-C. Chuang, R. J. Nicholas, K. S. Novoselov and A. K. Geim, Cyclotron resonance study of the electron and hole velocity in graphene monolayers. *Phys. Rev. B*, **76** (2007) 081406.

[16] Energies close to the Fermi energy are referred to as low energies. There is no standard definition of how close is *close enough*, but a range within ±1 eV is sometimes considered reasonable. In general, the range is application dependent.

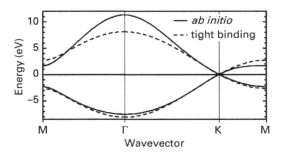

Fig. 3.7   Comparison of *ab-initio* and NNTB dispersions of graphene showing good agreement at low energies (energies about the K-point). $\gamma = 2.7$ eV and $s_0 = 0$ are used. Courtesy of S. Reich, J. Maultzsch and C. Thomsen, *Phys. Rev. B*, **66** (2002) 35412. Adapted with permission from S. Reich, J. Maultzsch and C. Thomsen, *Phys. Rev. B*, **66** (2002) 35412. Copyright (2002) by the American Physical Society.

where s_0 is called the overlap integral and is often employed as a fitting parameter with a value that is positive and nominally close to zero (compared with unity). The use of two fitting parameters (γ, s_0) will inevitably lead to a better overall agreement. Much of the exploration of graphene and derived nanostructures such as CNTs has been focused on the low-energy properties and dynamics; as such, we will use Eq. (3.37) in the remainder of our discussions unless noted otherwise.

Fermi energy

The equilibrium Fermi energy is the energy of the highest occupied k-state when the solid is in its ground or rest state (temperature 0 K). Determining E_F involves populating the k-states in the Brillouin zone with all the π-electrons in the solid according to Pauli's exclusion principle. There are N k-states in the valence band, which can hold $2N$ electrons, including spin degeneracy. Each carbon atom provides one p_z electron, resulting in two electrons/unit cell. Since there are N unit cells, we have a total of $2N$ electrons which will fill up the valence band. It follows that the highest occupied state housing the most energetic electrons are at the K-points, as identified earlier, and the corresponding energy is formally defined as the Fermi energy ($E_F = 0$ eV). The properties of electrons around the Fermi energy often determine the characteristics of practical electronic devices.

Figure 3.8 shows the 3D plot of the NNTB dispersion throughout the Brillouin zone. The upper half of the dispersion is the conduction (π^*) band and the lower half is the valence (π) band. Owing to the absence of a bandgap at the Fermi energy, and the fact that the conduction and valence bands touch at E_F, graphene is considered a semi-metal or zero-bandgap semiconductor, in contrast to a regular metal, where E_F is typically in the conduction band, and a regular semiconductor, where E_F is located inside a finite bandgap.

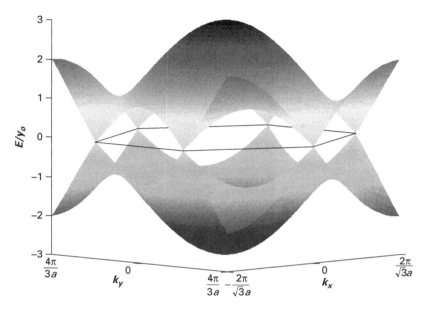

E/γ_o

$\dfrac{4\pi}{3a}$ k_y 0 $\dfrac{4\pi}{3a}$ $-\dfrac{2\pi}{\sqrt{3}a}$ k_x 0 $\dfrac{2\pi}{\sqrt{3}a}$

Fig. 3.8 The nearest neighbor tight-binding band structure of graphene. The hexagonal Brillouin zone is superimposed and touches the energy bands at the K-points.

Under non-equilibrium conditions (applied electric or magnetic fields) or extrinsic conditions (presence of impurity atoms), the Fermi energy will depart from its equilibrium value of 0 eV. The deviation of E_F from its equilibrium value is often valuable in determining the strength of the field or concentration of impurity atoms.

3.6 Linear energy dispersion and carrier density

The behavior of graphene electrons at the Fermi energy is of significant interest in condensed matter physics, particularly because the band structure has a linear dispersion which is representative of so-called massless particles (particles with zero effective mass).[17,18] For massless particles, Einstein's special relativity comes into play in the form of Dirac's relativistic quantum mechanical wave equation for describing the particle dynamics. Thus, the six K-points in graphene where the conduction and valence bands touch are frequently called the *Dirac points* (keep in mind that the energy at these points is, of course, E_F) in the physics literature.

[17] A linear relation between energy and momentum or wavevector formally defines a massless relativistic particle. Examples of massless (or approximately massless) relativistic particles include photons and neutrinos. The idea of effective mass will be formally introduced in Chapter 5.

[18] Massless electron waves also exist in a relatively new state of quantum matter called topological insulators. See Y. L. Chen *et al.*, Experimental realization of a three-dimensional topological insulator, Bi_2Te_3. *Science*, **325** (2009) 178–81.

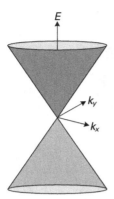

Fig. 3.9   The linear energy dispersion of graphene at the K-point which is known as the Dirac cone.

At these K-points and energies close by, the dispersion centered at the K-point can simply be expressed as a linear equation:

$$E(\mathbf{k})^{\pm}_{\text{linear}} = \pm \hbar v_{\text{F}} |\mathbf{k}| = \pm \hbar v_{\text{F}} \sqrt{k_x^2 + k_y^2} = \pm \hbar v_{\text{F}} k, \qquad (3.39)$$

where k is now in spherical coordinates, $\hbar$ is the reduced Planck's constant, and the Fermi velocity is defined as $v_{\text{F}} = (1/\hbar)(\partial E/\partial k)$ evaluated at the Fermi energy. It follows that $\hbar v_{\text{F}}$ is the gradient of the dispersion. A 3D plot of the linear dispersion is the celebrated *Dirac cone* shown in Figure 3.9. The linear dispersion has been confirmed up to approximately ± 0.6 eV by experimental spectroscopic measurements.[19]

A central property of electronic materials is the *density of states* (DOS) $g(E)$, which informs us of the density of mobile electrons or holes present in the solid at a given temperature. Formally, in two dimensions, the total number of states available between an energy E and an interval dE is given by the differential area in k-space dA divided by the area of one k-state. Mathematically, this is equivalent to

$$g(E)dE = 2g_z \frac{dA}{(2\pi)^2/\Omega}, \qquad (3.40)$$

where the factor of two in the numerator is included for spin degeneracy, g_z is the zone degeneracy, and Ω is the area of the lattice. There are six equivalent K-points, and each K-point is shared by three hexagons; therefore, $g_z = 2$ for graphene. To determine dA, let us consider a circle of constant energy in k-space. The perimeter of the circle is $2\pi k$ and the differential area obtained by an incremental increase

[19] I. Gierz, C. Riedl, U. Starke, C. R. Ast and K. Kern, Atomic hole doping of graphene. *Nano Lett.*, **8** (2008) 4603.

of the radius by dk is $2\pi k\,dk$.[20] Therefore, the DOS, which is always a positive value or zero, is

$$g(E) = \frac{2}{\pi}\left|k\frac{dk}{dE}\right| = \frac{2}{\pi}\left|k\left(\frac{dE}{dk}\right)^{-1}\right|, \tag{3.41}$$

where $g(E)$ has been normalized to the Ω. Substituting from Eq. (3.39) yields a linear DOS appropriate for low energies:

$$g(E) = \frac{2}{\pi(\hbar v_F)^2}|E| = \beta_g|E|, \tag{3.42}$$

where β_g is a material constant, $\beta_g \approx 1.5 \times 10^{14}\ \mathrm{eV^{-2}\,cm^{-2}} = 1.5 \times 10^6\ \mathrm{eV^{-2}\,\mu m^{-2}}$, and the absolute value of E is necessary because energy can be either positive (electrons) or negative (holes). At the Fermi energy ($E_F = 0$), the DOS vanishes to zero even though there is no bandgap. This is the reason why graphene is considered a semi-metal in contrast to regular metals that have a large DOS at the Fermi energy.

The electron carrier density is simply the number of states that are occupied per unit area at a given temperature. The occupation probability for electrons at finite temperatures is given by the *Fermi–Dirac* distribution:

$$f(E_F) = \frac{1}{1 + e^{(E-E_F)/k_B T}} \tag{3.43}$$

where k_B is Boltzmann's constant and T is the temperature. The equilibrium electron carrier density n is

$$n = \int_0^{E_{max}} g(E)f(E_F)\,dE = \frac{2}{\pi\hbar^2 v_F^2}\int_0^{E_{max}}\frac{E}{1 + e^{(E-E_F)/k_B T}}\,dE \tag{3.44}$$

where E_{max} is the maximum energy in the energy band. For the majority of applications of interest the Fermi energy is often much less than E_{max}, and owing to the exponential decay of the Fermi–Dirac distribution for energies greater than E_F, simply setting E_{max} to infinity introduces negligible error. Additionally, let $\eta = E_F/k_B T$ and $\eta_F = E_F/k_B T$ for mathematical convenience:

$$n = \frac{2}{\pi}\left(\frac{k_B T}{\hbar v_F}\right)^2\int_0^\infty \frac{\eta}{1 + e^{\eta-\eta_F}}\,d\eta = \frac{2}{\pi}\left(\frac{k_B T}{\hbar v_F}\right)^2 F_1(E_F/k_B T), \tag{3.45}$$

---

[20] The differential area can be visualized as the difference between the area of a circle of radius $k + dk$ and a circle of radius k in the limit when dk is very small.

with $F_1(\cdot)$ representing the Fermi–Dirac integral of order one.[21] In general, $F_1(\cdot)$ is an infinite series which is non-analytic but can be expressed in a closed form for a limited range of $E_F/k_B T$ by employing Taylor series expansion or other suitable approximation techniques. For the special case of an intrinsic graphene sheet with no doping of any kind, the Fermi energy is at 0 eV independent of temperature, resulting in an exact value of $\pi^2/12$ for the Fermi–Dirac integral. Therefore, the intrinsic carrier density n_i is

$$n_i = \frac{\pi}{6} \left(\frac{k_B T}{\hbar v_F} \right)^2 \approx 9 \times 10^5 T^2 \quad (\text{electrons/cm}^2), \tag{3.46}$$

with T in units of kelvin. It is worthwhile noting that the intrinsic carrier density reveals a temperature-squared dependence in contrast to conventional semiconductors, where n_i has an exponential dependence on temperature. Noticeably, the only material dependence is the Fermi velocity. At room temperature, $n_i \approx 8 \times 10^{10}$ cm^{-2}. The intrinsic carrier density is plotted in Figure 3.10a.

For an extrinsic material that has been doped with impurities for example, it is often desirable to have a simple closed-form equation for the extrinsic carrier density n_e as a function of E_F. Fortunately, an approximate algebraic formula is available by considering the limit when $E_F/k_B T \to \infty$. In this limit:

$$n_e \simeq \frac{\lambda}{\pi} \left(\frac{E_F}{\hbar v_F} \right)^2, \tag{3.47}$$

where λ is a fitting parameter. $\lambda = 1.1$ gives an error less than $\pm 10\,\%$ for $E_F/k_B T > 4$, and $\lambda = 1$ gives an error less than $5\,\%$ for $E_F/k_B T > 8$. This serves as a useful formula for extracting the Fermi energy for a given extrinsic doping concentration that is much greater than the intrinsic carrier density. The extrinsic carrier density is shown in Figure 3.10b.

Alternatively, a simple (approximate) formula that can be applied both for thermal equilibrium conditions and at finite potentials is obtained by the linear combination of the intrinsic and extrinsic expressions given by

$$n \cong n_i + \frac{E_F^2}{\pi (\hbar v_F)^2} \tag{3.48}$$

As a result of the electron–hole symmetry of the energy bands of graphene, the *hole carrier density p* under intrinsic conditions is equal to the electron carrier density,

---

[21] This *specific* Fermi–Dirac integral of order one is a form of alternating (infinite) power series that has a quasi-quadratic curve. For hand-analysis, it might be more convenient to express the Fermi–Dirac integral in terms of the polylogarithmic function, which is a standard mathematical function and whose properties are much more understood.

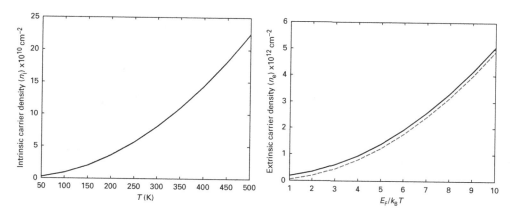

Fig. 3.10 (a) Intrinsic carrier density for graphene. (b) Extrinsic carrier density. The bold line represents the exact computation of Eq. (3.45) and the dashed line is the approximate closed-form formula of Eq. (3.47).

and for extrinsic conditions the hole carrier density has equivalent structure to the electron carrier density:

$$p = \frac{2}{\pi} \left(\frac{k_B T}{\hbar v_F} \right)^2 F_1(-E_F/k_B T). \tag{3.49}$$

3.7 Graphene nanoribbons

Graphene nanoribbons are narrow rectangles made from graphene sheets and have widths on the order of nanometers up to tens of nanometers. The nanoribbons can have arbitrarily long length and, as a result of their high aspect ratio, they are considered quasi-1D nanomaterials. GNRs are a relatively new class of nano-materials that can have metallic or semiconducting character, and are currently being investigated for their interesting electrical, optical, mechanical, thermal, and quantum-mechanical properties.[22]

There are two types of ideal GNR, which are called armchair GNRs (aGNRs) and zigzag GNRs (zGNRs). The aGNR has an armchair cross-section at the edges, while the zGNR has a zigzag cross-section, both illustrated in Figure 3.11. In addition, the GNRs are also labeled by the number of armchair or zigzag chains present in the width direction of the aGNR and zGNR respectively. Let N_a be the number of armchair chains and N_z the number of zigzag chains, then the nanoribbon can be conveniently denoted as N_a-aGNR and N_z-zGNR respectively. Figure 3.11 illustrates how to count the number of chains for a 9-aGNR and a 6-zGNR. The

---

[22] For a recent review of nano-scale and micrometer-scale graphene science, see A. K. Geim and K. S. Novoselov, The rise of graphene. *Nat. Mater.*, **6** (2007) 183–91.

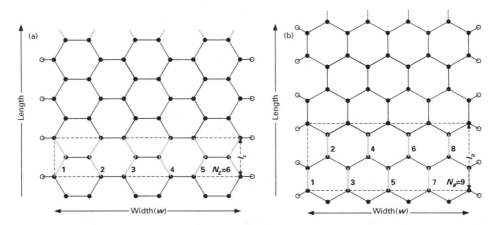

Fig. 3.11   The *finite-width* honeycomb structure of GNRs. (a) The lattice of a 6-zGNR and (b) the lattice of a 9-aGNR. The dashed box represents the primitive unit cell. The open circles at the edges denote passivation atoms such as hydrogen. The bold gray lines are the zigzag or armchair chains that are used to determine N_z or N_a respectively.

width of the GNRs can be expressed in terms of the number of lateral chains:

$$\text{aGNR}, w = \frac{N_a - 1}{2}a, \tag{3.50}$$

$$\text{zGNR}, w = \frac{3N_z - 2}{2\sqrt{3}}a, \tag{3.51}$$

where $a = 2.46$ Å is the graphene lattice constant as usual. The lengths of the primitive unit cells are $l_a = \sqrt{3}a$ and $l_z = a$ for aGNRs and zGNRs respectively. Non-ideal GNRs with mixed edge cross-sections may also exist, but those are not as well understood.

The small width of GNRs can lead to *quantum confinement* of electrons which restricts their motion to one dimension along the length of the nanoribbons, in contrast to a large graphene sheet where electrons are free to move in a 2D plane. As a result of several factors, including the quantum confinement, particular boundary conditions at the edges, and the effect of states arising from carbon atoms at the edges (also known as edge states), the band structure of GNRs is generally complex and departs significantly from that of the 2D graphene sheet. The band structure of GNRs can be computed numerically using first principles or tight-binding schemes.[23] The numerical computations reveal that zGNRs are semiconductors with bandgaps that are inversely proportional to the nanoribbons, width. Similarly, aGNRs also possess bandgaps which depend inversely on the width and, additionally, have a dependence on the number of armchair chains in

---

[23] See Y.-W. Son, M. L. Cohen and S. G. Louie, Energy gaps in graphene nanoribbons. *Phys. Rev. Lett.*, **97** (2006) 089901, and references contained therein. Currently, no accurate analytical formula for the band structure is available.

the nanoribbon. In both cases, the precise values of the bandgaps are sensitive to the passivation of the carbon atoms at the edges of the nanoribbons, although the general inverse width relation is preserved.

A useful first-order semi-empirical equation capturing the width dependence of the bandgap E_g has a simple relation:

$$E_g \approx \frac{\alpha}{w + w_o},\qquad(3.52)$$

where w (nm) is the width of the nanoribbons and w_o (nm) and α (eV nm) are fitting parameters. For zGNR, the fitting parameters can be considered constants, while for aGNR the fitting parameters have a dependence on N_a. Specifically, there are three types of aGNR resulting in three sets for w_o and α. The three types of aGNR are determined from whether $N_a = 3p$ or $N_a = 3p + 1$ or $N_a = 3p + 2$, where p is a positive integer. The values of the fitting parameters have been difficult to determine accurately from experimental data. This is primarily due to the challenges of accurately measuring the width and identifying the types of experimental nanoribbon. Experimentally extracted values of α range from 0.2 eV to about 1 eV.[24,25] Experimental and theoretical data suggest a $w_o \approx 1.5$ nm. As the width of the nanoribbon increases and exceeds about 50 or 100 nm, E_g vanishes and the band structure of GNRs gradually returns to that of a 2D graphene sheet. Figure 3.12 plots a set of experimentally extracted values for the bandgap confirming the inverse width dependence.

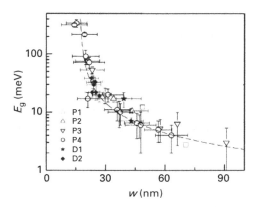

Fig. 3.12 Experimentally extracted bandgap versus width for GNRs. Adapted figure with permission from M. Han *et al.*[24] Copyright (2007) by the American Physical Society.

[24] M. Y. Han, B. Ozyiluaz, Y. Zhang and P. Kim, Energy band-gap engineering of graphene nanoribbons. *Phys. Rev. Lett.*, **98** (2007) 206805.
[25] X. Li, X. Wang, L. Zhang, S. Lee and H. Dai, Chemically derived, ultrasmooth graphene nanoribbon semiconductors. *Science*, **319** (2008) 1229–32.

3.8   Summary

The focus of this chapter has been on understanding the physical structure of graphene and developing a relatively simple theory for its electronic structure based on the p_z orbital nearest neighbor tight-binding formalism. In concluding the discussion on the electronic band structure of ideal 2D graphene it is important to emphasize that the NNTB dispersion of the π bands is primarily suitable for low energies, which covers most practical applications of interest. Of note is the electronic structure around the Fermi energy, which is often referred to as the Dirac cone, with much interesting physics, including insights about the conduct of relativistic particles. At high energies ($\gtrsim 3$ eV), it can be seen from Figure 3.6 that σ bands become dominant and, as a result, Eq. (3.37) cannot be expected to be useful. In addition, the NNTB developed here is for an ideally large (flat) sheet of graphene where edge effects and interactions with underlying substrates or overlying superstrates are negligible. The electronic structure of non-flat (i.e. warped or rippled) graphene sheets is a topic of contemporary scholarship.

Moreover, GNRs show a departure from the electronic properties of graphene sheets, most notably the opening of a bandgap due to the quantum confinement and edge effects. The opening of a bandgap is of great interest because it unlocks the potential of employing GNRs as transistors. It remains to be seen how controllable the synthesized bandgap is, and whether it can be repeatedly created in a routine manner. For this reason, GNRs are an active area of modern exploration with important implications for future applications.

3.9   Problem set

3.1. One-dimensional tight-binding model.

The goal of this exercise is to provide some mathematical fluency in the tight-binding formulation. Consider a 1D periodic arrangement of identical atoms with spacing a between the atoms. We will assume that every atom has one tightly bound electron. Note that this is a reasonable model for an electron in the s orbital. Furthermore, assume that the solutions for the atomic orbitals are known.

(a) Write down the expression for the (independent) electron wavefunction in the 1D solid.

(b) For this relatively simple case of one electron per unit cell, the allowed energies can be computed from the following formula:

$$E(k) = \frac{\int_{-\infty}^{\infty} \psi^* H \psi \, \mathrm{d}r}{\int_{-\infty}^{\infty} \psi^* \psi \, \mathrm{d}r}, \qquad (3.53)$$

which is the expectation value of the electron energy, which can also be interpreted as the average energy. Determine the energy dispersion within the nearest neighbor formalism and sketch the band structure.

3.2. A more accurate expression for the dispersion of electron waves in graphene.

(a) Show that for $S_{AB}(\mathbf{k}) \neq 0$ the energy dispersion is given by Eq. (3.38).

(b) For energies within ± 1 eV, compare Eq. (3.37) with Eq. (3.38) along the high-symmetry points Γ to K and K to M and quantify numerically the discrepancy between the two expressions.

Note that, in general, Eq. (3.38) is itself only an analytical approximation to the band structure computed via *ab-initio* methods (Figure 3.6); as such, this exercise is meant in part to provide an awareness of the additional discrepancies introduced by the electron–hole approximation ($S_{AB}(\mathbf{k}) = 0$), particularly at energies substantially removed from the Fermi level. Take $s_o = 0.05$ (or any other reasonable value of interest).

3.3. Electron–hole symmetry.

Show that, in the tight-binding formulation, electron–hole symmetry implies $S_{AB}(\mathbf{k}) = 0$ in the dispersion expression given in Eq. (3.22). For convenience, set $E_{2p} = 0$.

3.4. Derivation of the Dirac cone.

(a) Show that the $E-k$ dispersion of graphene is linear around the Dirac point by performing a first-order Taylor series expansion of the NNTB formula, Eq. (3.37).

(b) Determine the analytical expression for v_F. Measured estimates of v_F are around 10^6 m s^{-1}. Accordingly, what is the corresponding estimate of γ?

3.5. Relativistic massless particles.

(a) Perhaps the most fundamental physics about graphene is that the electrons behave like so-called *massless Dirac fermions*. Starting from Einstein's relativistic energy–momentum relation, show that, for a massless particle, Einstein's relation simplifies to a linear dispersion characteristic of graphene. (It is part of the exercise to recall Einstein's relation.) This implies in essence that electrons in graphene behave like relativistic particles.

(b) In the classical Newtonian model (kinetic energy–mass relation), what is the energy of a particle with vanishing mass?

3.6. Bandgap of GNRs.

Graphene nanoribbons are actively investigated today for a variety of applications, including nanoscale transistors. For reasons related to transistor leakage current, noise margin, and power dissipation, it is often desirable that the bandgap of the semiconductor be significantly larger than thermal energy, say bandgaps $\gtrsim 0.5$ eV are sought after. Assuming that one can estimate the bandgap fairly crudely (and conservatively) by $E_g \approx 0.9$ eV nm/w (nm) to within a factor of ~ 3.

(a) What nanoribbon width is required to obtain a bandgap of 0.5 eV?
(b) Comment on whether today's fabrication technology (optical or electron beam lithography) can achieve such dimensions?
(c) Survey the technical literature and identify two methods that have been employed to make nanoribbons with widths <10 nm.

4 Carbon nanotubes

Doing what others have not done demands a great deal of motivation.
Sumio Iijima (pioneer in the discovery and understanding of CNTs)

4.1 Introduction

There are two families of CNTs, namely single-wall CNTs and multi-wall CNTs (MWCNT) as shown in Figure 4.1. A single-wall CNT is a hollow cylindrical structure[1] of carbon atoms with a diameter that ranges from about 0.5 to 5 nm and lengths of the order of micrometers to centimeters.[2] An MWCNT is similar in structure to a single-wall CNT but has multiple nested or concentric cylindrical walls with the spacing between walls comparable to the interlayer spacing in graphite, approximately 0.34 nm. The ends of a CNT are often capped with a hemisphere of the buckyball structure. Carbon nanotubes are considered 1D nanomaterials owing to their very small diameter that confines electrons to move along their length. *The central goal of this chapter is to understand the physical structure of CNTs, and to determine their electronic band structures, which will enable us to gain insight into the properties and performance of CNT devices.*

The discussion in this chapter requires familiarity with the concepts developed in Chapters 2 and 3, such as crystal lattice and the band structure of graphene. Much of the content of this chapter is essential for the subsequent material presented throughout the book. We will use the acronym CNT to refer specifically to the single-wall variety unless explicitly stated otherwise. MWCNTs will be discussed mostly in the context of interconnect wires in Chapter 7.

This chapter begins by introducing the concept of chirality, which is the main idea used to describe the physical and electronic structure of CNTs. Then the physical structure of CNTs is elucidated conceptually as a folding operation of the graphene sheet resulting in three distinct configurations of single-wall CNTs.

---

[1] More precisely, a CNT has a polyhedral structure. However, for basic analysis and understanding, it is more convenient to consider nanotubes as cylinders.

[2] To put nanometers in practical perspective, 1 nm is ~50 000 times smaller than human hair (diameter of ~50 μm).

(a)

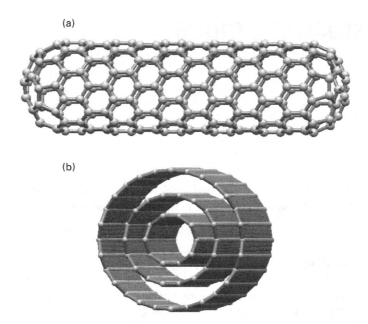

(b)

Fig. 4.1   Illustration of the two families of CNTs. (a) An ideal single-wall CNT with a hemispherical cap at both ends. (b) A MWCNT. In general, CNTs are much longer than depicted here, and the MWCNTs can have up to several dozen walls.

The chapter concludes by deriving the electronic band structure of CNTs from the band structure of graphene.

4.2   Chirality: a concept to describe nanotubes

Chirality is the key concept used to identify and describe the different configurations of CNTs and their resulting electronic band structure. Since the concept of chirality is of fundamental importance and often unfamiliar to engineers,[3] let us take a moment to introduce the concept of chirality before discussing how it is applied to describe CNT structure. The term chirality is derived from the Greek term for hand, and it is used to describe the reflection symmetry between an object and its mirror image.[4] Formally, *a chiral object is an object that is not superimposable on its mirror image*; and conversely, *an achiral object is an object that is superimposable on its mirror image*. At this point, a visual illustration is often

---

[3] Chirality (or handedness) was defined by Lord Kelvin in 1884 to describe reflection symmetry of molecules. It is a widely used concept in chemistry, particle physics, and in the study of symmetry groups in the mathematical sciences.

[4] Care must be taken when discussing the mirror image of an object as it relates to chirality. For the example of the left hand, the plane containing the hand should be normal to the mirror.

invaluable in making these definitions vividly clear. For example, consider the left hand; its mirror image is the right hand and we find that it is not possible to superimpose the two hands or images such that all the features coincide precisely, as illustrated in Figure 4.2. Therefore, the human hand is a chiral object. Now, consider a circle as another example; its mirror image is also an identical circle which superimposes precisely on top of the original image. Therefore, a circle is an

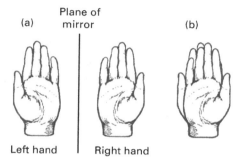

Fig. 4.2   Example of a chiral object. (a) The left hand and its mirror image (right hand). (b) It is not possible to superimpose the left hand on the right hand; therefore, the human hand is chiral.

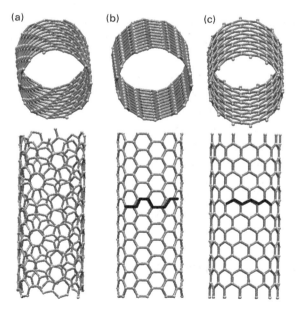

Fig. 4.3   The three types of single-wall CNT: (a) A chiral CNT, (b) an armchair CNT, and (c) a zigzag CNT. The cross-sections of the latter two illustrations have been highlighted by the bold lines showing the armchair and zigzag character respectively.

achiral object. In a general usage, chirality is invoked to highlight the presence or lack of mirror symmetry that provides intuition about understanding phenomena.

Understanding the concept of chirality is essential, because it is used to classify the physical and electronic structure of CNTs. The CNTs that are superimposable on their own mirror images are classified as *achiral CNTs*, and all other nanotubes that are not superimposable are classified as *chiral CNTs*. To be clearer, all single-wall CNTs are either chiral or achiral. Moreover, achiral CNTs are further classified as *armchair CNTs* or *zigzag CNTs*, depending on the geometry of the nanotube circular cross-section. To summarize briefly, there are three types of single-wall CNT: chiral CNTs, armchair CNTs, and zigzag CNTs, of which the latter two are achiral and their symmetry often makes them easier to explore and gain broad insight. The three types of single-wall CNT and their associated geometrical cross-sections are shown in Figure 4.3. Indeed, it is a worthwhile educational exercise for the reader to examine the structure of all three CNTs and confirm the chiral or achiral properties. Better yet, the reader is encouraged to purchase a *ball-and-stick* chemistry model set and actually construct CNTs to gain hands-on familiarity with the physical structure.

4.3   The CNT lattice

We introduced the concept of chirality to classify the different types of CNT in the previous section, but it was not at all clear why CNTs arrange to form a chiral or an achiral geometry. Fortunately, it is actually fairly easy to understand the origin of the different types of CNT by considering that a CNT results from folding or wrapping of a graphene sheet. To see how the folding operation works, we start from the direct lattice of graphene and then define a mathematical construction which folds graphene's lattice into a CNT. Moreover, this mathematical folding construction directly leads to a precise determination of the primitive lattice of carbon nanotubes, which is required information in order to derive the CNT band structure. It is very important to keep in mind that the folding of graphene to form a CNT is simply a convenient conceptual idea to study the basic properties of CNTs. In actuality, CNTs naturally grow as a cylindrical structure, often with the aid of a catalyst, which does not involve folding of graphene in any physical sense.

Figure 4.4a shows the honeycomb lattice of graphene and the primitive lattice vectors $\mathbf{a}_1$ and $\mathbf{a}_2$, defined on a plane with unit vectors $\hat{\mathbf{x}}$ and $\hat{\mathbf{y}}$:

$$\mathbf{a}_1 = \left[\frac{\sqrt{3}a}{2}, \frac{a}{2} \right], \quad \mathbf{a}_2 = \left[\frac{\sqrt{3}a}{2}, -\frac{a}{2} \right], \tag{4.1}$$

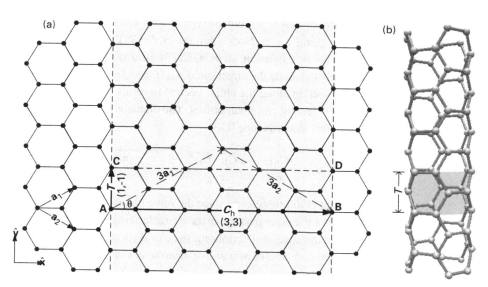

Fig. 4.4 An illustration to describe the conceptual construction of a CNT from graphene. (a) Wrapping or folding the dashed line containing points A and C to the dashed line containing points B and D results in the (3, 3) armchair carbon nanotube in (b) with $\theta = 30°$. The CNT primitive unit cell is the cylinder formed by wrapping line AC onto BD and is also highlighted in (b).

where a is the underlying Bravais lattice constant, $a = \sqrt{3}a_{C-C} = 2.46\,\text{Å}$, and a_{C-C} is the carbon–carbon bond length (~ 1.42 Å). Also, $\mathbf{a}_1 \cdot \mathbf{a}_1 = \mathbf{a}_2 \cdot \mathbf{a}_2 = a^2$, $\mathbf{a}_1 \cdot \mathbf{a}_2 = a^2/2$, and the angle between $\mathbf{a}_1$ and $\mathbf{a}_2$ is 60°. With reference to Figure 4.4a, a single-wall CNT can be conceptually conceived by considering folding the dashed line containing primitive lattice points A and C with the dashed line containing primitive lattice points B and D such that point A coincides with B, and C with D to form the nanotube shown in Figure 4.4b. The CNT is characterized by three geometrical parameters, the chiral vector $\boldsymbol{C}_h$,[5] the translation vector $\boldsymbol{T}$, and the chiral angle θ, as shown in Figure 4.4a. The chiral vector is the geometrical parameter that uniquely defines a CNT, and $|\boldsymbol{C}_h| = C_h$ is the CNT circumference. C_h is defined as the vector connecting any two primitive lattice points of graphene such that when folded into a nanotube these two points are coincidental or indistinguishable. For the particular exercise of Figure 4.4, the chiral vector is the vector from point A to B, $\boldsymbol{C}_h = 3\mathbf{a}_1 + 3\mathbf{a}_2 = (3,3)$. In general:

$$\boldsymbol{C}_h = n\mathbf{a}_1 + m\mathbf{a}_2 = (n, m), (n, m \text{ are positive integers}, 0 \leq m \leq n) \qquad (4.2)$$

and the resulting carbon nanotube is described as an (n, m) CNT.

---

[5] It is perhaps more enlightening to think of the chiral vector as a roll-up, wrapping, or folding vector, because it describes the roll-up of graphene into a nanotube. Some authors also refer to it as the circumferential vector.

Important observations regarding the type of CNT can be deduced directly from the values of the chiral vector. Notice that the $(3, 3)$ CNT of Figure 4.4 leads to an armchair nanotube. By extension, all (n, n) CNTs are armchair nanotubes. The case when C_h is purely the along the direction of $\mathbf{a}_1$, $(C_h = (n, 0))$ can be visually seen (from the cross-section along the chiral vector) to result in zigzag nanotubes. All other (n, m) CNTs lead to chiral nanotubes. The diameter d_t of a carbon nanotube is derived from its circumference $|C_h|$:

$$d_t = \frac{|C_h|}{\pi} = \frac{\sqrt{C_h \cdot C_h}}{\pi} = \frac{a\sqrt{n^2 + nm + m^2}}{\pi}. \tag{4.3}$$

Notably, different chiralities can produce the same nanotube diameter; as a result, the diameter is not a unique parameter for characterizing CNTs. To see this more clearly, consider the case of determining the chiral (or armchair) nanotubes that have the same exact diameter as a zigzag nanotube. For this exercise, we differentiate between zigzag and chiral nanotubes by using $(n_z, 0)$ and (n, m) to refer to zigzag and chiral CNTs respectively. In order to produce the same diameter, the condition

$$n_z^2 = n^2 + nm + m^2 \tag{4.4}$$

must be satisfied. The CNTs that satisfy this condition are more easily seen in Figure 4.5, which is a contour plot of the RHS of Eq. (4.4), and the lines are the constant-diameter lines from zigzag nanotubes (LHS of Eq. (4.4)). To interpret the constant-diameter plot, let us determine the chiral nanotube that has exactly the same diameter as the $(19, 0)$ CNT. By following the contour line connected to $n = 19, m = 0$ in Figure 4.5, we identify that $n = 16, m = 5$ also (precisely) intersects the line. To sum up, $(19, 0)$ and $(16, 5)$ CNTs produce the same diameter. The constant-diameter plot is also useful because it provides us with the chiral indices of nanotubes that have diameters close to a particular zigzag CNT. The equivalence of the diameter among dissimilar nanotubes has important implications for the electronic properties, as is shown for example in the derivation of the bandgap in Section 4.8.

The other two geometrical parameters (T and θ) can be derived from the chiral vector. For instance, the chiral angle is the angle between the chiral vector and the primitive lattice vector $\mathbf{a}_1$:

$$\cos\theta = \frac{C_h \cdot \mathbf{a}_1}{|C_h||\mathbf{a}_1|} = \frac{2n + m}{2\sqrt{n^2 + nm + m^2}}. \tag{4.5}$$

The chiral angle can be viewed as describing the tilt angle of the hexagons relative to the tubular axis. Owing to the sixfold hexagonal symmetry of the honeycomb

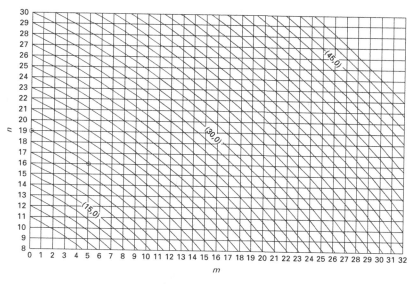

Fig. 4.5 Constant-diameter contour plot of Eq. (4.4). The contour lines represent the constant diameter of an $(n, 0)$ CNT, and some contour lines have been labeled with their corresponding $(n, 0)$ index for convenience. As an example, the circles show that a $(19, 0)$ CNT has identical diameter to a $(16, 5)$ nanotube.

lattice, unique values of the chiral angle are restricted to $0 \leq \theta \leq 30°$. For the particular exercise of Figure 4.4, $\theta = 30°$. In general, all armchair nanotubes have a chiral angle of $30°$, and $\theta = 0°$ for all zigzag nanotubes.

In order to determine the primitive unit cell of the CNT, we need to consider the translation vector which defines the periodicity of the lattice along the tubular axis. Geometrically, T is the smallest graphene lattice vector perpendicular to C_h. As can be seen from Figure 4.4, $T = (1, -1)$ for all armchair nanotubes. Similarly, the translation vector for all zigzag nanotubes can be visually deduced to be $T = (1, -2)$. More broadly, the translation vector can be computed from the orthogonality condition $C_h \cdot T = 0$. Let $T = t_1 \mathbf{a}_1 + t_2 \mathbf{a}_2$, where t_1 and t_2 are integers. Therefore:

$$C_h \cdot T = t_1(2n + m) + t_2(2m + n) = 0. \qquad (4.6)$$

Determining the acceptable solution for t_1 and t_2 requires a subtle interplay involving mathematical analysis and visual insight. There are two orthogonal directions ($\pm 90°$) relating T to C_h, and solving for either direction leads to an equivalent solution for the translation vector. Let us restrict the direction to $+90°$ as shown in Figure 4.4a. Then, according to the orientation definition of the lattice vectors $\mathbf{a}_1$ and $\mathbf{a}_2$, t_1 must be a positive integer and t_2 must be a negative integer for T to be

$+90°$ with respect to C_h. With this visual insight, one set of integers that satisfy Eq. (4.6) is $(t_1, t_2) = (2m + n, -2n - m)$. However, deeper thinking reveals that there are several sets of integers that are also solutions of Eq. (4.6). For instance, consider an $(8, 2)$ CNT; $(t_1, t_2) = (12, -18)$ is a solution, but so are $(t_1, t_2) = (12, -18)/2$, $(t_1, t_2) = (12, -18)/3$, and $(t_1, t_2) = (12, -18)/6$. The actual *acceptable* solution that leads to the shortest translation vector is $(t_1, t_2) = (12, -18)/6 = (2, -3)$, where the factor of 6 is the greatest common divisor of 12 and 18. Hence, the acceptable solution for Eq. (4.6) is

$$T = (t_1, t_2) = \left(\frac{2m + n}{g_d}, -\frac{2n + m}{g_d} \right), \qquad (4.7)$$

where g_d is the greatest common divisor of $2m + n$ and $2n + m$. The length of the translation vector is

$$|T| = T = \frac{\sqrt{3}|C_h|}{g_d} = \frac{\sqrt{3}\pi d_t}{g_d}. \qquad (4.8)$$

The chiral and translation vectors define the primitive unit cell of the CNT, which is a cylinder with diameter d_t and length T. Some auxiliary results that are useful to compute include the surface area of the CNT unit cell, the number of hexagons per unit cell, and the number of carbon atoms per unit cell. The surface area of the CNT primitive unit cell is the area of the rectangle defined by the C_h and T vectors, $|C_h \times T|$. The number of hexagons per unit cell N is the surface area divided by the area of one hexagon:

$$N = \frac{|C_h \times T|}{|a_1 \times a_2|} = \frac{2(n^2 + nm + m^2)}{g_d} = \frac{2|C_h|^2}{a^2 g_d}. \qquad (4.9)$$

This simplifies to $N = 2n$ for both armchair and zigzag nanotubes. Since there are two carbon atoms per hexagon, there are a total of $2N$ carbon atoms in each CNT unit cell. A summary of the geometric parameters and associated equations for CNTs nanotubes are listed in Table 4.1. Specific values of the geometric parameters for selected nanotubes ranging in diameter from 1 to 3 nm are shown in Table 4.2.

In order to gain hands-on familiarity with the conceptual construction of a CNT, the reader is encouraged to construct a nanotube from the blank graphene sheet in Figure 4.14 at the end of this chapter. As an example, the reader can construct a $(4, 1)$ CNT and conveniently verify it with the construction shown in Figure 4.6. For the full construction experience, the reader should physically fold the coincident points in the lattice onto each other to create a paper model of the CNT.

Table 4.1. Table of parameters and associated equations for CNTs.[a]

Symbol	Name	cCNT	aCNT	zCNT
C_h	chiral vector	$C_h = n\mathbf{a}_1 + m\mathbf{a}_2 = (n, m)$	$C_h = (n, n)$	$C_h = (n, 0)$
C_h	length of chiral vector	$C_h = |C_h| = a\sqrt{n^2 + nm + m^2}$	$C_h = a\sqrt{3}n$	$C_h = an$
d_t	diameter	$d_t = \frac{a}{\pi}\sqrt{n^2 + nm + m^2}$	$d_t = \frac{an}{\pi}\sqrt{3}$	$d_t = \frac{an}{\pi}$
θ	chiral angle	$\cos\theta = \frac{2n+m}{2\sqrt{n^2+nm+m^2}}$	$\theta = 30°$	$\theta = 0°$
g_d	greatest common divisor	$g_d \equiv \gcd(2m+n, 2n+m)$	$g_d = 3n$	$g_d = n$
T	translation vector	$T = \frac{2m+n}{g_d}\mathbf{a}_1 - \frac{2n+m}{g_d}\mathbf{a}_2$	$T = \mathbf{a}_1 - \mathbf{a}_2$	$T = \mathbf{a}_1 - 2\mathbf{a}_2$
T	length of translation vector	$T = |T| = \frac{\sqrt{3}C_h}{g_d}$	$T = a$	$T = a\sqrt{3}$
N	number of hexagons/cell	$N = \frac{2C_h^2}{a^2 g_d}$	$N = 2n$	$N = 2n$

[a] The primitive basis vectors $\mathbf{a}_1$ and $\mathbf{a}_2$ are defined according to Eq. (4.1). cCNT stands for chiral CNT, aCNT for armchair CNT, and zCNT for zigzag CNT.

Table 4.2. Table of specific values for selected CNTs of diameters ~1–3 nm.[a]

C_h	d_t (nm)	C_h (nm)	T (nm)	Θ (deg)	N	E_g (eV)
(10, 4)	0.98	3.07	0.89	16.1	52	0
(10, 5)	1.04	3.25	1.13	19.1	70	0.86
(13, 0)	1.02	3.20	0.43	0	26	0.84
(15, 15)	2.03	6.40	0.25	30	30	0
(16, 5)	1.49	4.67	8.10	13.2	722	0.60
(16, 14)	2.04	6.40	5.54	27.8	676	0.43
(19, 0)	1.49	4.67	0.43	0	38	0.58
(26, 0)	2.04	6.40	0.43	0	52	0.44
(32, 0)	2.51	7.87	0.43	0	64	0.35
(38, 0)	2.98	9.35	0.43	0	76	0.30

[a] The bandgap E_g is computed from the tight-binding band structure of CNTs, which is discussed in Sections 4.6 and 4.8.

4.4 CNT Brillouin zone

Given the primitive unit cell of CNTs developed in the previous section, we are now in a position to construct the CNT reciprocal lattice and Brillouin zone which will subsequently aid us in determining its electronic band structure. The focus will mostly be on the first Brillouin zone, which contains the unique values of the allowed wavevectors and energies. In a way analogous to the path taken in the prior section to construct the CNT physical structure from the honeycomb lattice of graphene, we will discover that the Brillouin zone of CNTs is composed of a

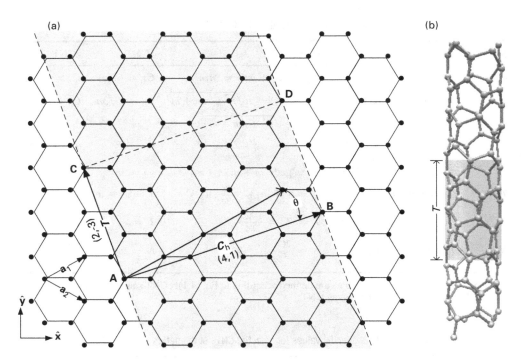

Fig. 4.6   Construction of a chiral CNT. (a) A (4, 1) CNT is constructed from the lattice of graphene. Points A, B, C, and D are used in a similar manner as in Figure 4.4. (b) Cylindrical structure of the (4, 1) CNT.

series of cross-sections or line cuts of the reciprocal lattice of graphene. The basis vectors for graphene's reciprocal lattice are

$$\mathbf{b}_1 = \left(\frac{2\pi}{\sqrt{3}a}, \frac{2\pi}{a} \right), \quad \mathbf{b}_2 = \left(\frac{2\pi}{\sqrt{3}a}, -\frac{2\pi}{a} \right). \tag{4.10}$$

The wavevectors defining the CNT of the first Brillouin zone are the reciprocals of the primitive unit cell vectors given by the reciprocity condition previously discussed in Chapter 2 (Eq. (2.31)):

$$e^{i(\mathbf{K}_a + \mathbf{K}_c) \cdot (\mathbf{C}_h + \mathbf{T})} = 1, \tag{4.11}$$

where $\mathbf{K}_a$ is the reciprocal lattice vector along the nanotube axis and $\mathbf{K}_c$ is along the circumferential direction, both given in terms of the reciprocal lattice basis vectors of graphene ($\mathbf{b}_1$, $\mathbf{b}_2$). Equation (4.11) simplifies to

$$\mathbf{C}_h \cdot \mathbf{K}_c = 2\pi, \quad \mathbf{T} \cdot \mathbf{K}_c = 0, \tag{4.12}$$

$$\mathbf{C}_h \cdot \mathbf{K}_a = 0, \quad \mathbf{T} \cdot \mathbf{K}_a = 2\pi. \tag{4.13}$$

Employing the expressions for C_h, T, and N in Table 4.1, the wavevectors can be derived algebraically:

$$K_a = \frac{1}{N}(m\mathbf{b}_1 - n\mathbf{b}_2), \tag{4.14}$$

$$K_c = \frac{1}{N}(-t_2\mathbf{b}_1 + t_1\mathbf{b}_2). \tag{4.15}$$

The lengths of the reciprocal lattice wavevectors are inversely proportional to the CNT lattice dimensions, i.e. $|K_a| = 2\pi/T$ and $|K_c| = 2\pi/C_h$. K_a and K_c in a sense describe the nanotube Brillouin zone.

The next step is to determine the allowed wavevectors *within* the Brillouin zone that lead to Bloch wave functions. Let us consider a nanotube of length $L_t = N_{uc}T$ where N_{uc} is the number of CNT unit cells in the nanotube. The allowed wavevectors k along the axial direction are obtained from the periodic boundary conditions on the Bloch wave functions:

$$\psi(0) = \psi(L_t) = e^{ikN_{uc}T}\psi(0), \quad \Rightarrow \therefore e^{ikN_{uc}T} = 1, \tag{4.16}$$

resulting in the set of wavevectors

$$k = \frac{2\pi}{N_{uc}T}l, \quad l = 0, 1, \ldots, N_{uc} - 1, \tag{4.17}$$

where the maximum integer value of l is determined from the requirement that unique solutions for k are restricted to the first Brillouin zone, i.e. maximum $(k) < |K_a| = 2\pi/T$.[6] In the limit where the CNT is very long, for instance $L_t \gg T$ or $N_{uc} \gg 1$,[7] then the spacing between k-values vanishes and, to first-order, k can be considered a continuous variable along the axial direction:

$$k = \left(-\frac{\pi}{T}, \frac{\pi}{T}\right), \tag{4.18}$$

where the wavevector has been re-centered to be symmetric about zero consistent with standard Brillouin zone convention.

Applying the same periodic boundary conditions to determine the allowed wavevectors q along the circumferential direction yields

$$\psi(0) = \psi(C_h) = e^{iqC_h}\psi(0), \quad \Rightarrow \therefore e^{iqC_h} = 1, \tag{4.19}$$

$$q = \frac{2\pi}{C_h}j = \frac{2}{d_t}j = j|K_c|, \quad j = 0, 1, \ldots, j_{max}. \tag{4.20}$$

---

[6] Recall from Chapter 2 that unique values for k and energy are *always* contained within the first Brillouin zone. Any k-value outside the first Brillouin zone can be mathematically translated back into the first Brillouin zone by a reciprocal lattice vector.

[7] This requirement is often satisfied by practical CNTs. Furthermore, theoretical work has shown that a continuously varying k remains a fairly reasonable approximation for CNTs as short as 10 nm. See A. Rochefort, D. R. Salahub and P. Avouris, Effects of finite length on the electronic structure of carbon nanotubes. *J. Phys. Chem. B*, **103**, (1999) 641–6.

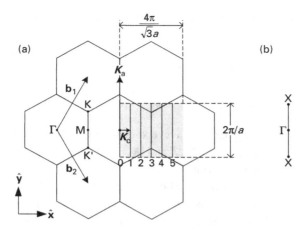

Fig. 4.7   Brillouin zone of a (3, 3) armchair CNT (shaded rectangle) overlaid on the reciprocal lattice of graphene. The numbers refer to $j = 0, 1, \ldots, 5$ for a total of $N = 6$ 1D bands in the CNT Brillouin zone. The central hexagon is the first Brillouin zone of graphene, and the high-symmetry points (Γ, M, and K) of graphene's Brillouin zone are also indicated.[8] The area of the CNT Brillouin zone is equal to the area of graphene's Brillouin zone. (b) The high-symmetry points of a line representing a CNT 1D band is illustrated.

We observe that the q-values are separated by a gap that is much greater than the spacing in k-values, i.e. $2\pi/C_h \gg 2\pi/L_t$ for long CNTs with lengths $L_t \gg C_h$. Therefore, the q variable is quantized or discretely spaced compared with the relatively continuous k variable, which implies that the allowed CNT wavevectors in the Brillouin zone are composed of a series of lines as shown in Figure 4.7a. These lines are basically 1D cuts of graphene's reciprocal lattice. The final question we have to resolve to obtain the complete set of 1D lines is the maximum value of j in Eq. (4.20) that yields the total set of unique values for q. To answer this question we deduce that since the unique wavevectors are discrete set of line cuts of graphene's reciprocal lattice, then any two line cuts or q-values that are separated by a reciprocal lattice vector of graphene must be equivalent. The shortest reciprocal lattice vector of graphene that is an integer multiple K_c is NK_c.[9] As a result, the maximum value of q is less than $N|K_c|$, and hence

$$q = \frac{2\pi}{|C_h|}j, \quad j = 0, 1, \ldots, N - 1. \tag{4.21}$$

[8] We noted in Chapter 3 that the K'-point is essentially equivalent to the K-point except under certain inquiries. A case in point is during conservation of momentum in interband electron scattering in CNTs (more about this in Chapter 7).

[9] K_c given by Eq. (4.15) is a CNT reciprocal lattice vector but not a graphene reciprocal lattice vector. To obtain a reciprocal lattice vector that is common to both CNT and graphene requires multiplying K_c by N, $NK_c = -t_2\mathbf{b}_1 + t_1\mathbf{b}_2$.

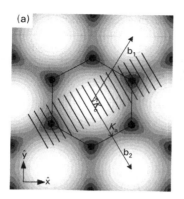
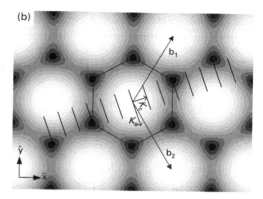

Fig. 4.8   Brillouin zone of (a) (10, 0) CNT and (b) a (4, 1) CNT, overlaid on the contour plot of the conduction band of graphene (darker shades corresponds to lower energies). The Brillouin zone of CNTs consists of the series of dark lines representing the N (20 and 14 respectively) 1D bands. The Brillouin zone of graphene is the hexagon. Note that the N lines have been folded to be symmetric with graphene's hexagonal Brillouin zone for convenience (the unfolded lines only exist in the $+K_c$ direction). The lengths of the lines are $|K_a| = 2\pi/T$.

Figure 4.8 shows the Brillouin zones for the (10, 0) and (4, 1) CNTs. The area of the Brillouin zone of a nanotube is equal to the area of graphene's Brillouin zone $(8\pi^2/\sqrt{3}a^2)$, which is a consequence of the fact that CNT 1D wavevectors are cuts of graphene's 2D Brillouin zone. Table 4.3 is a summary of the expressions for the reciprocal lattice vectors and wavevectors defined in this section.

We now strive to combine expressions for the allowed axial and circumferential Brillouin zone wavevectors (Eqs. (4.18) and (4.21) respectively) in order to generate an expression for any arbitrary allowed state or wavevector within the Brillouin zone. This general Brillouin zone wavevector **k** is what will be used to compute the allowed energies in the band structure of CNTs and is given by

$$\mathbf{k} = k\frac{K_a}{|K_a|} + q\frac{K_c}{|K_c|}, \tag{4.22}$$

where $K_a/|K_a|$ and $K_c/|K_c|$ are the unit vectors in the axial and circumferential directions respectively. Substituting for $|K_a|$, $|K_c|$, and q from Table 4.3 yields

$$\mathbf{k} = k\frac{K_a}{2\pi/T} + jK_c, \quad \left(j = 0, 1, \ldots, N-1, \text{ and } -\frac{\pi}{T} < k < \frac{\pi}{T}\right). \tag{4.23}$$

In summary, each value of j corresponds to a line or 1D band with wave vectors k ranging from $-\pi/T$ to $+\pi/T$. This is one of the most important results the reader should appreciate.

Table 4.3. Summary of CNT reciprocal lattice vectors and Brillouin zone wavevectors.[a]

Symbol	Name	cCNT	aCNT	zCNT
K_c	circumferential lattice vector	$K_c = \frac{(m+2n)\mathbf{b}_1+(n+2m)\mathbf{b}_2}{2(n^2+nm+m^2)}$	$K_c = \frac{\mathbf{b}_1+\mathbf{b}_2}{2n}$	$K_c = \frac{2\mathbf{b}_1+\mathbf{b}_2}{2n}$
$|K_c|$	length of K_c	$|K_c| = \frac{2\pi}{C_h}$	$|K_c| = \frac{2\pi}{\sqrt{3}an}$	$|K_c| = \frac{2\pi}{an}$
K_a	axial lattice vector	$K_a = \frac{m\mathbf{b}_1-\mathbf{b}_2}{N}$	$K_a = \frac{\mathbf{b}_1-\mathbf{b}_2}{2}$	$K_a = -\frac{\mathbf{b}_2}{2}$
$|K_a|$	length of K_a	$|K_a| = \frac{2\pi}{T}$	$|K_a| = \frac{2\pi}{a}$	$|K_a| = \frac{2\pi}{\sqrt{3}a}$
k	axial Brillouin zone wavevector	$k = \left(-\frac{\pi}{T}, \frac{\pi}{T}\right)$	$k = \left(-\frac{\pi}{a}, \frac{\pi}{a}\right)$	$k = \left(-\frac{\pi}{\sqrt{3}a}, \frac{\pi}{\sqrt{3}a}\right)$
q	circumferential Brillouin zone wavevector	$q = \frac{2\pi}{C_h}j$	$q = \frac{2\pi}{\sqrt{3}an}j$	$q = \frac{2\pi}{an}j$
k	Brillouin zone wavevector	$\mathbf{k} = k\frac{K_a}{|K_a|} + q\frac{K_c}{|K_c|}$	$\mathbf{k} = \frac{k}{2\pi/a}K_a + jK_c$	$\mathbf{k} = \frac{k}{2\pi/\sqrt{3}a}K_a + jK_c$

[a] The primitive basis vectors $\mathbf{b}_1$ and $\mathbf{b}_2$ are defined according to Eq. (4.10). j is an integer from 0 to $N-1$.

4.5   General observations from the Brillouin zone

In the previous section we learned that the finite width C_h of CNTs leads to quantization of the CNT Brillouin zone, which essentially results in the CNT Brillouin zone being composed of a series of 1D cuts of the reciprocal lattice of graphene. This implies that the CNT band structure will be 1D cross-sections of the band structure of graphene. Before actually computing the CNT band structure, we can arrive at very important broad conclusions from insights gained from leveraging our understanding of the electronic structure of graphene.

Let us consider an armchair and zigzag CNT to get our intuition running, and then generalize to a chiral nanotube. We recall from Chapter 2 that the conduction and valence bands of graphene touch at the K-points, where the highest equilibrium occupied states (corresponding to the Fermi energy) exist. The degeneracy or touching of the bands at a K-point resulted in the absence of a bandgap which explains the metallic behavior of graphene. At every other point in the Brillouin zone of graphene there exists an energy gap between the conduction and valence bands. We can then expect that if any of the CNT 1D bands or Brillouin zone lines cuts the reciprocal lattice of graphene at a K-point, then the nanotube will be metallic, otherwise the nanotube will have gaps between conduction and valence bands and, hence, be semiconducting. For example, the fourth band of the (3, 3) armchair nanotube intersects two hexagonal corner points of graphene (see K and

K′ in Figure 4.7a) leading to the conclusion that the $(3, 3)$ CNT is metallic. Indeed, it is straightforward to show that arbitrary (n, n) armchair CNTs are metallic. To derive this, let us recall the vector from the Γ-point to the other high-symmetry points of the hexagonal Brillouin zone of graphene:

$$\Gamma M = \frac{\mathbf{b}_1 + \mathbf{b}_2}{2} = \left(\frac{2\pi}{\sqrt{3}a}, 0\right), \quad \Gamma K = \left(\frac{2\pi}{\sqrt{3}a}, \frac{2\pi}{3a}\right). \tag{4.24}$$

The length of the 1D bands of an CNT is $2\pi/a$, which is greater than the lengths of the sides of the hexagonal Brillouin zone of graphene (length $= 4\pi/3a$). As a result, if any of the 1D bands of armchair nanotubes intersect an M-point of the hexagon, it will also simultaneously intersect a K-point. Therefore, in order for a 1D band of an armchair CNT to be metallic, the ΓM vector has to be an integer multiple of $\mathbf{K}_c$. Mathematically, this condition is equivalent to[10]

$$j|\mathbf{K}_c| = j\frac{2\pi}{\sqrt{3}an} \equiv |\Gamma M|, \quad \therefore j = n. \tag{4.25}$$

This condition is satisfied by all armchair nanotubes by the $j = n$th band or Brillouin zone line at $k = \pm 2\pi/3a$. Hence, in general, armchair CNTs are metallic.

Likewise, we can apply similar reasoning to zigzag nanotubes to determine the conditions in which they are metallic. The Brillouin zone of a $(10, 0)$ zigzag CNT was previously shown in Figure 4.8a revealing that the 1D bands are parallel to the ΓK vector. Hence, for a zigzag nanotube to be metallic, the ΓK vector has to be an integer multiple of $\mathbf{K}_c$:

$$j|\mathbf{K}_c| = j\frac{2\pi}{an} \equiv |\Gamma K|, \quad \therefore j = \frac{2}{3}n, \tag{4.26}$$

which is satisfied when $j = 2n/3$ at $k = 0$. However, since j is restricted to integer values (see Eq. (4.21)), only a zigzag CNT with a chirality that is an integer multiple of 3 (i.e. $n/3$ is an integer) is metallic, otherwise the zigzag CNT is semiconducting. For example, $(12, 0)$, $(15, 0)$, $(18, 0)$ are metallic CNTs, whereas $(10, 0)$, $(11, 0)$, and $(13, 0)$ are semiconducting nanotubes. In general, an arbitrary chiral CNT is metallic if the angle between $j\mathbf{K}_c$ and the ΓK vector is the chiral angle:

$$j|\mathbf{K}_c| = j\frac{2\pi}{a\sqrt{n^2 + nm + m^2}} \equiv |\Gamma K|\cos\theta = \frac{2\pi(2n + m)}{3a\sqrt{n^2 + nm + m^2}}, \tag{4.27}$$

which is satisfied only when $j = (2n + m)/3$. This leads to the celebrated condition that a CNT is metallic if $(2n + m)$ or equivalently $(n - m)$ is an integer multiple of 3 or 0,[11] otherwise the CNT is semiconducting. Invariably, a very important

---

[10] The equivalence symbol $\equiv$ is used to enforce the equivalence of the LHS expression and the RHS expression.

[11] $j = (2n + m)/3$ is equivalent to $j = [(n - m)/3] + [(n + 2m)/3]$, which results in an integer value for j when $n - m$ is a multiple of 3.

question that naturally arises is: What are the percentages of metallic and semiconducting CNTs based on a random (non-preferential) chirality distribution? We can compute these percentages by considering the sum of all equally probable chirality combinations up to (n, n) that are an integer multiple of 3:

$$S_m = \sum_{i=1}^{n} \sum_{m=0}^{i} H\left[-\text{mod}\left(\frac{n-m}{3}\right)\right], \tag{4.28}$$

where $H(\cdot)$ is the Heaviside step function and $\text{mod}(n - m/3)$ gives the remainder of $(n - m)$ divided by 3, which can either be 0, 1, or 2. The Heaviside function is used to discretize the result: $H(\cdot) = 1$ if $(n - m)$ is an integer multiple of 3 or $H(\cdot) = 0$ otherwise. S_m is the sum of all the metallic CNTs in the combination. The total number of chirality combinations up to (n, n) is given by the sum of the linear arithmetic series (S_{tot}).

$$S_{tot} = \sum_{i=1}^{n} \sum_{m=0}^{i} 1 = \frac{n(n+3)}{2}. \tag{4.29}$$

Therefore, the probability that a CNT is metallic is S_m/S_{tot}, and the probability it is semiconducting is $1 - (S_m/S_{tot})$. For large values of the chiral index n, say $n > 100$, the probability of metallic and semiconducting nanotubes approaches 1/3 and 2/3 respectively.[12]

Also noteworthy is the existence of band degeneracy, i.e. some of the lines or CNT 1D bands are cuts of equivalent regions of the reciprocal lattice of graphene. For example, there are two lines that touch equivalent M-points in Figure 4.8a. Similarly, there are two lines that touch two K-points in Figure 4.8b. In both of these examples the lines are double degenerate. We will further highlight degeneracy in the CNT band structure computation discussed in subsequent sections; and as a prelude, degenerate bands have identical energy dispersion.

4.6   Tight-binding dispersion of chiral nanotubes

The band structure of CNTs can be determined from the NNTB energy dispersion of graphene. This is sometimes referred to as *zone folding*, because the energy bands of CNTs are line cuts or cross-sections of the bands of graphene. It follows that the entire Brillouin zone of CNTs can be folded into the first Brillouin zone of graphene. The zone-folding technique is a powerful yet simple method to determine

---

[12] This statistical result should be considered a *rule of thumb* for random chiralities. In experimental synthesis of CNTs, depending on the specific details, certain chiralities might be more (energetically or kinetically) favored to grow over other chiralities and, as a result, the *rule of thumb* might not apply.

the electronic properties of CNTs and the performance of CNT devices. However, the zone-folding and NNTB method is limited, as it does not account for several phenomena which are particularly pronounced for small diameter ($d_t < 1$ nm) nanotubes and high-energy excitations, and will be discussed further in Section 4.9. As such, the NNTB band structure is primarily useful for CNTs with $d_t > 1$ nm operating at low energies,[13] which fortunately covers the majority of electronic and sensor applications of CNTs.

It is now timely to introduce the high-symmetry points of CNTs to aid us in discussing their band properties. High-symmetry points are specific functions of geometry. In the case of graphene, the Brillouin zone has a hexagonal geometry and there are three points of symmetry: Γ, M, and K (see Figure 4.7a). For CNTs, the Brillouin zone is composed of N lines. By convention, the high-symmetry points of a line are the center and end of the line, which are labeled Γ and X respectively. It follows that each line will have a Γ-point and two X-points (see Figure 4.7b).

The band structure of CNTs can be computed by inserting the allowed wavevectors into the energy dispersion relation for graphene. We recall from Chapter 3 the NNTB energy E dispersion relation of graphene:

$$E(\mathbf{k})^{\pm} = \pm\gamma\sqrt{1 + 4\cos\left(\frac{\sqrt{3}a}{2}k_x\right)\cos\left(\frac{a}{2}k_y\right) + 4\cos^2\left(\frac{a}{2}k_y\right)}, \quad (4.30)$$

where the plus and minus signs refer to the conduction and valence bands respectively and where $\gamma \sim 3.1$ eV will be employed unless stated otherwise. $\mathbf{k}$ will now refer to the CNT arbitrary Brillouin zone wavevector given by Eq. (4.23), and can be rewritten in terms of its $\hat{x}$ and $\hat{y}$ components as $\mathbf{k} = k_x\hat{x} + k_y\hat{y}$. Expanding Eq. (4.23) into the $\hat{x}$, $\hat{y}$ coordinates gives the vector components for arbitrary chiral CNTs:

$$k_x = \frac{2\pi\sqrt{3}aj(n+m)C_h + a^3k(n^3 - m^3)}{2C_h^3}, \quad (4.31)$$

$$k_y = \frac{\sqrt{3}ak(n+m)C_h + 2\pi aj(n-m)}{2C_h^2}, \quad (4.32)$$

where k varies according to Eq. (4.17) and j takes on discrete values from 0 to $N-1$. In general, for any (n, m) CNT, there will be N valence bands ($E \leq 0$) and N conduction bands ($E \geq 0$). Each one of the bands has $2N_{uc}$ allowed states, where the factor of 2 is due to spin degeneracy. Hence, there are a total number of $2N_{uc}N$ available states each in the valence and conduction parts of the band structure. Since there are $2N$ electrons in the unit cell of a CNT and N_{uc} unit cells, it follows

---

[13] The phrase *low energies* will be used frequently, and refers to energies not far away from the Fermi energy.

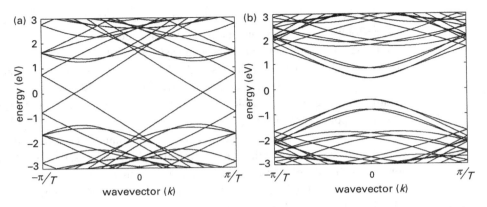

Fig. 4.9 Band structures for (a) (10, 4) metallic CNT and (b) (10, 5) semiconducting CNT, within ± 3 eV. The CNT diameters are 0.98 nm and 1.04 nm respectively. The metallic CNT shows a band degeneracy at 0 eV and $k = \pm 2\pi/3T$. The semiconducting CNT has a bandgap of ~ 0.86 eV.

that $2N_{uc}N$ electrons need to be accommodated. Invariably, at equilibrium the valence bands will be fully occupied and the conduction bands empty with the Fermi energy $E_F = 0$ eV. Figure 4.9. shows the band structures for (10, 4) metallic and (10, 5) semiconducting chiral CNTs. The semiconducting (10, 5) CNT has a bandgap $E_g \sim 0.86$ eV at the Γ-point. We will show later in Section 4.8 that the bandgap is inversely proportional to the diameter, $E_g \sim 0.9$ (nm eV)$/d_t$, where d_t is in nanometers.

In the subsequent sections we will explore in more detail the band structure of the highly symmetric achiral nanotubes to elucidate general properties of metallic and semiconducting CNTs and introduce useful approximations to describe the lowest energy bands which are of greatest interest.

4.7   Band structure of armchair nanotubes

The Brillouin zone wavevector (Eq. (4.23)) for armchair CNTs expressed in the $\hat{x}$ and $\hat{y}$ coordinates is $\mathbf{k} = (2\pi j/\sqrt{3}an)\hat{x} + k\hat{y}$. Substituting into Eq. (4.30) yields the energy dispersion E_{ac} for armchair nanotubes.

$$E_{ac}(j, k) = \pm \gamma \sqrt{1 + 4\cos\left(\frac{j\pi}{n}\right)\cos\left(\frac{ka}{2}\right) + 4\cos^2\left(\frac{ka}{2}\right)},$$

$$\left(j = 0, 1, \ldots, 2n - 1, \text{ and } -\frac{\pi}{a} < k < \frac{\pi}{a}\right). \tag{4.33}$$

The band structure for an (8, 8) armchair nanotube is shown in Figure 4.10, revealing an energy degeneracy at $ka = \pm 2\pi/3$, where the valence band touches the

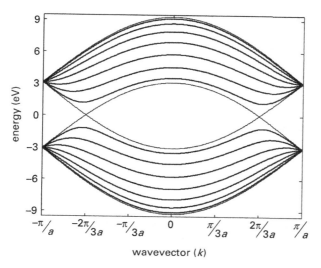

wavevector (k)

Fig. 4.10   Band structure for (8, 8) armchair nanotube containing 16 1D subbands in the valence and conduction bands each. For all armchair CNTs, the valence band touches the conduction band at $ka = \pm 2\pi/3$, which explains their metallic properties. The thin lines are for the non-degenerate subbands, while the thick lines are for doubly degenerate subbands.

conduction band. In general, the energy degeneracy at 0 eV is common to all armchair CNTs and, hence, armchair CNTs are metallic. Additionally, the lowest and highest energy subbands[14] of the valence and conduction bands are non-degenerate at arbitrary k-values with all other subbands having a twofold degeneracy. Noticeably, all the subbands have a large degeneracy of $2n$ at the zone edge ($ka = \pm\pi$) corresponding to $E_{ac} = \pm\gamma$.

A particularly important observation is that the first subbands of the valence and conduction bands have a linear dispersion at low energies and to a good approximation can be approximated in a simple manner with a linear $E-k$ relation independent of chirality. The linear dispersion for the right-half of the Brillouin zone can be expressed as

$$E_{\text{linear}}(k)^{\pm} \approx \pm\hbar v_F \left| k - \frac{2\pi}{3a} \right|, \quad \left(\frac{\pi}{3a} < k < \frac{\pi}{a} \right), \tag{4.34}$$

where k has a range $(2\pi/3a)$ that is restricted to prevent overrunning the edge of the Brillouin zone. $\hbar$ is the reduced Planck's constant and v_F is the velocity at the Fermi energy (also known as the Fermi velocity). The Fermi velocity will be more formally discussed in the next chapter, which elaborates on the equilibrium properties of CNTs. In short, the Fermi velocity can be shown to be $v_F = (1/\hbar)(\partial E/\partial k)$,

---

[14] The term subband will now be used routinely to denote one of the N 1D bands in the band structure of CNTs. Of central interest is the first subband, which refers to the lowest (highest) energy subband of the conduction (valence) band.

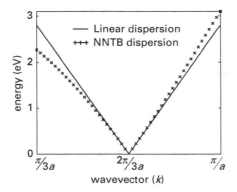

Fig. 4.11   Comparison between the linear dispersion and the tight-binding dispersion for the first subband of the conduction band of armchair nanotubes with good accuracy ($|$error$| < 7\%$ at 1 eV).

and it follows that the linear dispersion has a slope with magnitude $\sim h v_F$. The actual NNTB dispersion for the first subband obtained by substituting $j = n$ into Eq. (4.33) is $E_{ac}(k) = +\gamma|1 - 2\cos(ka/2)|$, independent of the value of n, and the linear dispersion is simply its first-order Taylor expansion. Comparison of the linear dispersion approximation with the actual tight-binding dispersion for an (8, 8) armchair nanotube is shown in Figure 4.11, showing that the linear dispersion is a good approximation for energies up to about 1 eV. Owing to mirror symmetry of the Brillouin zone, the linear dispersion of the left-half portion can be obtained from Eq. (4.34) by imposing the restriction $-\pi/a < k < -\pi/3a$ and interchanging $-2\pi/3$ with $+2\pi/3$.

Another practically useful result, especially in electron transport, is the energies of the band minima of higher subbands (e.g. second, third subbands). These energies are especially relevant in determining the contribution of higher subbands to electron transport in large-diameter nanotubes ($d_t \gtrsim 3$ nm), as will be discussed in Chapter 6. The subband minimums $E_{ac,j}$ in the vicinity of the Fermi energy are roughly located at $k \sim \pm 2\pi/3a$, which reduces Eq. (4.33) to

$$E_{ac,j} \approx \gamma \sqrt{2 + 2\cos\left(\frac{j\pi}{n}\right)}. \tag{4.35}$$

A first-order Taylor series approximation about $j = n$ provides a simpler expression:

$$E_{ac,j} \approx \frac{\gamma\pi}{n}|j - n|. \tag{4.36}$$

The first-order Taylor series approximation is, in fact, rather accurate for the lowest subbands because the energy dispersion about the Fermi energy is largely linear.

For the first-subband, $j = n$ and we obtain the expected result of 0 eV. The absolute sign reflects the fact that the next higher subbands have a twofold degeneracy. That is, for example, $j = n + 1$ and $j = n - 1$ produce the same energy.

4.8 Band structure of zigzag nanotubes and the derivation of the bandgap

Zigzag CNTs are perhaps the most attractive type of nanotube to explore because of the presence of either metallic or semiconducting behavior. They also possess high symmetry, leading to simple analytical expressions for many of the solid-state properties. The energy dispersion of zigzag CNTs can be obtained from the Brillouin zone wavevector (Eq. (4.23)), which reduces to

$$\mathbf{k} = \frac{2\pi\sqrt{3}j - nka}{2an}\hat{\mathbf{x}} + \frac{2\pi j + \sqrt{3}nka}{2an}\hat{\mathbf{y}}. \tag{4.37}$$

Substituting into Eq. (4.30) yields the energy dispersion E_{zz} for zigzag nanotubes:

$$E_{zz}(j, k) = \pm\gamma\sqrt{1 + 4\cos\left(\frac{\sqrt{3}ka}{2}\right)\cos\left(\frac{j\pi}{n}\right) + 4\cos^2\left(\frac{j\pi}{n}\right)},$$

$$\left(j = 0, 1, \ldots, 2n - 1, \text{ and } -\frac{\pi}{\sqrt{3}a} < k < \frac{\pi}{\sqrt{3}a}\right). \tag{4.38}$$

The band structures for the metallic $(12, 0)$ and semiconducting $(13, 0)$ nanotubes are shown in Figure 4.12. In general, when n is a multiple of 3, the zigzag CNT

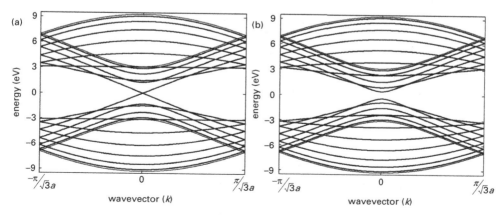

Fig. 4.12   Band structures for (a) $(12, 0)$ and (b) $(13, 0)$ CNTs. The $(12, 0)$ CNT is metallic, while the $(13, 0)$ CNT is semiconducting due to the bandgap at $k = 0$. The thin lines indicate non-degenerate subbands, while the thick lines are for doubly degenerate subbands.

is metallic, otherwise it is semiconducting. The lowest subbands have a twofold degeneracy for both metallic and semiconducting zigzag nanotubes. For the metallic case, a simple linear $E - k$ relation can accurately describe the first subband of the valence and conduction bands. Similar to graphene's linear dispersion, the linear dispersion for the first subband ($j = 2n/3$ with a twofold degeneracy) of metallic zigzag CNTs can be expressed as

$$E_{\text{linear}}(k)^{\pm} \approx \pm h v_{\text{F}} |k|. \tag{4.39}$$

In semiconductor theory, the bandgap E_g is of fundamental importance in determining its solid-state properties and also electronic transport in device applications. The band index for the first subband (j_1) for semiconducting zigzag nanotubes is $2n/3$ rounded to the nearest integer, which expressed mathematically is

$$j_1 = \text{round}\left(\frac{2n}{3}\right) \approx \frac{2}{3}n + \frac{1}{3}, \tag{4.40}$$

where round $(\cdot)$ is a function that converts its argument to the nearest integer and the RHS expression is a linear approximation to the staircase-like round function. The linear approximation is actually exact for semiconducting zigzag nanotubes when $n - 1$ is an integer multiple of 3. Substituting the linear approximation for j into Eq. (4.38), the bandgap for semiconducting zigzag CNTs is

$$E_g \approx 2\gamma \left| 1 + 2\cos\left(\frac{2\pi}{3} + \frac{\pi}{3n}\right) \right|. \tag{4.41}$$

For relatively large n, $\pi/3n$ is a small perturbation around $2\pi/3$; therefore, a first-order Taylor series expansion of Eq. (4.41) about $2\pi/3$ in the cosine argument leads to

$$E_g \approx 2\gamma \left(\frac{2\pi n + \pi}{\sqrt{3}n} - \frac{2\pi}{\sqrt{3}}\right) = 2\gamma \frac{\pi}{\sqrt{3}n}. \tag{4.42}$$

The chiral index is related to the CNT diameter as shown in Table 4.1 for zigzag CNTs, thereby simplifying Eq. (4.42) to

$$E_g \approx 2\gamma \frac{a_{\text{C-C}}}{d_{\text{t}}}, \tag{4.43}$$

which is the frequently employed bandgap–diameter relation for CNTs. Numerically, $E_g(\text{eV}) \sim 0.9/d_{\text{t}}$ (nm), a useful formula for quickly estimating a nanotube's bandgap in electron-volts. Figure 4.13 demonstrates the accuracy of Eq. (4.43) compared with exact NNTB bandgap computation. The figure includes the bandgaps of all semiconducting CNTs with chiral indices ranging from $(7, 0)$ to $(29, 28)$. Remarkably, even though Eq. (4.43) was derived for zigzag

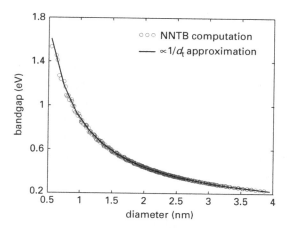

Fig. 4.13
The bandgap of semiconducting CNTs calculated using Eq. (4.43) (solid line) compared with exact NNTB computation showing good agreement. This plot includes all semiconducting CNTs with chiral indices ranging from (7, 0) to (29, 28). The bandgap predictions for nanotubes with diameters < 1 nm should be considered crude estimates only because the NNTB computation is inaccurate for sub-nanometer CNTs owing to the large curvature.

semiconducting CNTs, it is equally accurate in estimating the bandgaps of arbitrary chiral nanotubes. This case in point serves to illustrate the utility of the highly symmetric zigzag nanotubes as an excellent vehicle for exploring the general properties of arbitrary chiral nanotubes. It also illustrates that the bandgaps of semiconducting nanotubes are primarily dependent on the CNT diameter, not the details of their chirality.

4.9 Limitations of the tight-binding formalism

The NNTB formalism is a simple yet powerful analytical technique in describing the electronic structure of π electrons in graphene and CNTs. Its accuracy is largely judged by how well it reproduces *ab-initio* or first-principles band structure calculations. It is particularly most accurate for low-energy electrons in CNTs with diameters >1 nm, which covers the majority of electronics-based applications. The limitation of the tight-binding formalism becomes increasingly pronounced when considering small diameter (less than ~0.7–1 nm) nanotubes and high-energy electron excitations. Users of the tight-binding band structure would be well advised to verify that it is applicable to the CNTs of their interest, especially for either sub-nanometer diameters or high-energy operation. At the very least one should be aware of potential shortcomings of the tight-binding predictions. This section enumerates on some of the main limitations of the tight-binding band structure.

(a) *Electron–hole symmetry at high energies*: At energies increasingly higher than the Fermi energy, electron–hole symmetry gradually fades away, as is evident in the *ab-initio* band structure of graphene shown in Chapter 3. To account properly for the lack of electron–hole symmetry, the overlap fitting parameter s_0 should be finite and positive with a value that is nominally close to zero compared with unity (see Eq. (3.38)). As a rule of thumb, electron–hole symmetry should be invoked with care for subbands greater than the first or second subband in the band structure of CNTs. For much higher energies than the Fermi energy, inclusion of the overlap fitting parameter might not suffice. In that case, up to third nearest neighbors in the tight-binding formalism might be needed for band structure accuracy.[15]

(b) *Sigma electrons at high energies*: In the tight-binding formalism, our concern has been with the relatively delocalized π electrons that are the mobile electrons in the material. At sufficiently high energies (approximately $\pm > 3$ eV), σ electrons from the sigma bonds between carbon atoms becoming increasingly mobile and, therefore, lead to new energy–wavevector branches in the band structure of graphene (see Figure 3.6) and, consequently, CNTs. The effect of σ electrons should be taken into consideration (through *ab initio* or comparable computations) in special cases including high-energy photon excitations.

(c) *Curvature effects in small nanotubes*: Curvature effects refer to a collection of phenomena that become pronounced in small-diameter CNTs (diameters <1 nm) owing to their large curvature. The interesting phenomena include carbon–carbon bond length (a_{C-C}) asymmetry and dependence on the curvature, and $\sigma - \pi$ orbital overlap and hybridization. For small-diameter nanotubes, the carbon–carbon bond length along the circumference of the nanotube is somewhat stretched due to the large curvature compared with carbon–carbon bond length along the axial direction. This bond length asymmetry results in a slight shift of the K-point of graphene further along the y-axis (see Figure 4.7a for the Brillouin zone coordinates) with the major outcome that otherwise metallic zigzag and chiral nanotubes now acquire a small bandgap and are a widely referred to as quasi-metallic nanotubes. Armchair nanotubes still preserve their metallic character even in the presence of the large curvature, since their 1D bands are entirely along the y-axis. Theoretical and experimental results have shown that $E_g \sim 1/d_t^2$ in quasi-metallic CNTs.[16]

Additionally, the large curvature warps the orbitals such that the π orbitals are not truly orthogonal to the σ orbitals due to the curved space leading to

[15] S. Reich, J. Maultzsch, C. Thomsen and P. Ordejon, Tight-binding description of graphene. *Phys. Rev. B*, **66** (2002) 035412.

[16] A. Kleiner and S. Eggert, Curvature, hybridization, and STM images of carbon nanotubes. *Phys. Rev. B*, **64**, (2001) 113402. O. Gülseren, T. Yildirim and S. Ciraci, Systematic *ab initio* study of curvature effects in carbon nanotubes. *Phys. Rev. B*, **65** (2002) 153405.

hybridization of the σ and π orbitals. The net effect of the $\sigma - \pi$ hybridization includes bandgap adjustments and fairly complex band structure modifications that become increasingly pronounced at energies further away from the Fermi energy.[17] Hybridization effects might play a substantial role in applications that exploit optical or photon transitions in CNTs.

(d) *Substrate effects*: The CNTs considered so far have been pristine nanotubes with no supporting material. However, practical nanotubes are often on top of a substrate wafer and the presence of the wafer and topography of the wafer surface might lead to small deformations in the CNT geometry or long-range electrostatic forces from the potential energy in the substrate material. For example, certain substrates, such as crystalline quartz, play a role in aligning the growth of CNTs exclusively along a specific orientation on the substrate. These crystalline substrates may also introduce a non-negligible periodic potential along the axial direction of the nanotube that needs to be considered in the Hamiltonian, which might lead to noticeable band structure modifications. Periodic substrate effects are not well understood at the moment, and should be kept in mind when the substrate is observed or suspected of playing a strong role in CNT growth or device operation.

4.10 Summary

The physical and electronic structure of CNTs has been elucidated in this chapter. One of the overriding themes observed throughout the chapter is the concept of symmetry. For example, nanotubes are classified based on the symmetry of their physical structure, resulting in armchair, zigzag, and chiral nanotubes, where the armchair and zigzag types enjoy a higher symmetry than chiral nanotubes. The high symmetry of achiral nanotubes is of practical convenience, particularly for analysis and insight owing to the simple expressions for the energy dispersions. Fundamentally, CNTs can be understood as a folding or wrapping of a graphene sheet. As such, both their physical and electronic properties are derived from graphene. Invariably, a good understanding of the physical and electronic structure of graphene is required in order to fully appreciate the behavior of electrons in CNTs. It is worthwhile noting that viewing nanotubes as a folded graphene sheet is one way to understand their properties. It is also entirely possible to consider nanotubes directly as a cylindrical structure and determine their Bravais lattice and Brillouin zone ultimately leading to their band structure within the tight-binding formalism. Perhaps such an alternative approach might be less efficient in producing insight and information about the electronic structure.

---

[17] Curvature effects are discussed in greater detail in Chapter 3 of S. Reich, C. Thomsen and J. Maultzsch *Carbon Nanotubes: Basic Concepts and Physical Properties* by (Wiley-VCH, 2004).

Electronically, CNTs can be either metallic or semiconducting, and this diversity makes CNTs very attractive for a wide variety of applications, including interconnects, transistors, and sensors. The electronic structure of nanotubes has been understood mostly from a relatively simple nearest-neighbor π-orbital tight-binding model, which has so far proved to be particularly useful in describing the low-energy behavior of charge carriers in nanotube devices with diameters greater than about 1 nm. A key property of the band structure is the horizontal and vertical mirror symmetry. The vertical mirror symmetry, also known as electron–hole symmetry, holds within low-energy excitation. It remains to be firmly determined what the threshold energy is that distinguishes between low-energy and high-energy operation. The successful analytical development of the dispersion of electrons in nanotubes cannot be overemphasized, and is the foundation for much of the working theory of nanotube electronic properties and device behavior.

4.11   Problem set

All the problems are intended to exercise and refine analytical techniques while providing important insights. If a particular problem is not clear, the reader is encouraged to re-study the appropriate sections. Some problems might involve making reasonable approximations or assumptions beyond what is already stated in the specific question in order to obtain a final answer. The reader should not consider this frustrating, because this is obviously how problems are solved in the real world.

4.1. Construction of a (5, 0) CNT.

One way to get intimately familiar with the structure of CNTs is actually to construct one. Ideally, it would be lovely to have a bunch of balls acting as carbon atoms and some sticks pretending to be the bonds and arrange the balls and sticks in a polyhedral cylindrical manner to build a CNT. For the purpose of this exercise, we will have to make do with a paper construction of a CNT using Figure 4.14.

(a) Construct a (5, 0) zigzag CNT (similar to Figure 4.4a). Show the lattice basis vectors, CNT unit cell, and the chiral and translation vectors.
(b) Fold or wrap the appropriate points on the paper to create a paper model of the nanotube.
(c) What is the diameter and surface area of the unit cell of the (5, 0) CNT?
(d) What is the bandgap of the nanotube?
(e) Sketch the Brillouin zone of the (5, 0) CNT overlaid on the reciprocal lattice of graphene.
(f) How many subbands are non-degenerate at arbitrary wavevectors? That is, how many subbands have identical energy and wavevector values in

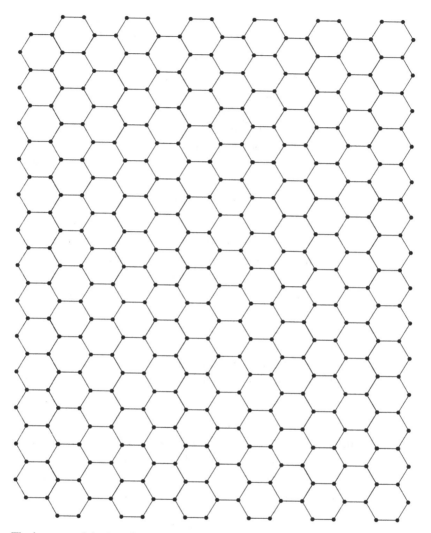

Fig. 4.14   The honeycomb lattice of graphene.

the first Brillouin zone? (*Hint*: a shortcut to the answer is available by inspection of the Brillouin zone.)

4.2. Construction of a (5, 5) CNT.
   Redo Problems 4.1 for a (5, 5) CNT.

4.3. Starting from Eq. (4.11), derive the expressions for K_a and K_c as shown in Eqs. (4.14) and (4.15).

4.4. Chiral index convention.
   By convention, the geometric description of CNTs is restricted to chiral indices from $(n, 0)$ to (n, m), where n and m are integers and $m \leq n$. The

immediate question then is: Why is $m > n$ not considered interesting? To address this question, show that an $(n, n+p)$ CNT is identical to an $(n+p, n)$ CNT where p is a positive integer.

4.5. Carbon nanotube subband degeneracies.

Accurately keeping track of the CNT subband degeneracies (subbands with identical $E - k$ relation) is routinely necessary even for the most basic nanotube analysis. While tracking degeneracies should be somewhat trivial, it is quite common for new readers to be puzzled by a factor of two (or more) every now and then. The purpose of this exercise is to give the reader some familiarity with the degeneracies that occur, particularly at the high-symmetry points of the nanotube band structure.

(a) What are the band degeneracies for the CNT subband that is at the Γ-point and K-point of the Brillouin zone of graphene? That is, a subband that touches these symmetry points. Determine an answer by direct analysis from the Brillouin zone of graphene for armchair and zigzag nanotubes.

(b) Likewise, what are the band degeneracies for the CNT subband in the interior of the Brillouin zone of graphene? Determine an answer by direct analysis from the Brillouin zone of graphene for armchair and zigzag chiral nanotubes.

The emphasis on the direct analysis of the Brillouin zone of graphene is because this is the fastest and easiest way to determine nanotube subband degeneracies.

4.6. Carbon nanotube second lowest subband.

In the modeling of electron transport in CNTs, it is generally desirable to include contributions to current from the lowest and higher subbands. In many (but not all) cases, only the two lowest subbands will effectively determine charge transport. Therefore, an accurate analytical expression for the band index and energy minima of the second subband will be most welcome. Consider a semiconducting zigzag nanotube.

(a) Derive an approximate analytical expression for the second subband index and the subband energy minima.

(b) Determine the accuracy of the energy minima from part (a) by comparing with the tight-binding predictions for zigzag nanotubes with diameters from 1 to 3 nm.

(c) What is the approximate expression for the energy difference between the first subband and the second subband?

4.7. Robustness of random distribution of the metallic/semiconducting character of nanotubes.

It is commonly known and shown in the text that a random distribution of large numbers of CNTs will reveal that one-third are metallic and the remaining two-third semiconducting. However, one cannot help but wonder if this is also the case for a limited sample of CNTs. Show that the 1/3 ratio is fairly robust even for a limited range of random chiralities. For example, in practice, grown CNTs are usually within the 0.7 to 2 nm range. Let us say we examine chiral indices from (8, 0) to (30, 0) combinations to represent the range of diameters of interest.

(a) What is the probability of metallic and semiconducting nanotubes within this range of chiral indices? Quantify the departure (if any, in percent) from the 1/3, 2/3 probabilities expected for large numbers of nanotubes.

(b) Imagine there was a method to produce only chiral nanotubes. What will be the resulting probabilities of metallic and semiconducting CNTs?

4.8. Probability of zigzag CNTs.

Zigzag CNTs are usually easier to model and analyze owing to their high-symmetry structure that affords simpler closed-form mathematics. In many cases, the general properties of semiconducting CNTs are then inferred by analyzing a semiconducting zigzag nanotube of comparable diameter.

(a) One question of intellectual curiosity then arises: What is the probability that a CNT is a semiconducting zigzag CNT from a random distribution of a large number of carbon nanotubes?

(b) Compare the probability of a semiconducting zigzag nanotube with that of the 2/3 expected for an arbitrary semiconducting nanotube. Which type of nanotube (zigzag or chiral) makes up the lion's share of the 2/3 probability of semiconducting nanotubes?

(c) Out of a large ensemble of randomly distributed zigzag nanotubes, what is the probability of metallic and semiconducting zigzag nanotubes in the ensemble?

4.9. From nanotubes to nanoribbons.

It has recently been shown that GNRs can be produced by *unzipping* CNTs along the longitudinal direction.

(a) What type of GNR is expected if a zigzag nanotube is unzipped?

(b) What type of GNR is expected if an armchair nanotube is unzipped?

5   Carbon nanotube equilibrium properties

5.1   Introduction

This chapter explores the equilibrium and thermodynamic electronic properties of single-wall CNTs. Equilibrium refers to the state of a system in the absence of external forces, and thermodynamics accounts for the evolution of the macroscopic properties of the system with temperature. The system we are referring to here is of course CNTs, and the properties of interest include the DOS, group velocity, effective mass, and charge carrier density. These properties are of central importance for understanding and predicting the electrical, optical, and thermal behavior of CNTs. Our learning path will utilize the development of analytical expressions of these properties to provide insight and understanding regarding the inherent solid-state behavior. Additionally, analytical expressions are especially desirable for compact modeling of nanotube devices.

In order to be comfortable with the discussion of the equilibrium properties, the reader should be familiar with the band structure of CNTs developed in the previous chapter. *In fact, it will be worthwhile to have in hand a copy of all the band structure figures shown in Chapter 4 as one goes through this chapter.* The equilibrium properties will be employed repeatedly in subsequent chapters to describe transport in CNTs under the quasi-equilibrium assumption applicable at low energies. We begin by discussing the DOS of the free-electron gas in 1D space. This provides perspective, allowing us to appreciate and relate to the actual DOS in CNTs which are quasi-1D solids. In general, the DOS is of fundamental importance in understanding crystalline solids, and many of the other equilibrium properties can be derived from this important parameter. We conclude the chapter by deriving the electron and hole carrier densities at finite temperatures.

5.2 Free-electron density of states in one dimension

The DOS is a fundamental property of solids that describes the number of states available to be occupied by a particle at every allowed energy. The particles could be elementary wave-like particles, such as electrons, holes, photons, and phonons, or composite wave-like entities, such as excitons (electron–hole pairs). The particles of primary interest to us are electrons and holes. In a casual sense a state is akin to a home; and just like a home can house say one or two persons, a state can also accommodate one or two particles. In the same sense, the DOS in a solid is analogous to the density of homes in a city. It follows, then, that the number of particles present in a solid can be calculated in a straightforward manner by estimating the number of states that are occupied. Mathematically, the DOS $g(E)$ (where E is the particle energy) for a 1D solid can be interpreted based on the form that it is presented, such as

$$g(E) \Rightarrow \text{DOS per unit length at energy } E,$$

$$g(E)\, dE \Rightarrow \text{total number of states per unit length between } E \text{ and } E + dE,$$

$$\int_{E_1}^{E_2} g(E)\, dE \Rightarrow \text{total number of states per unit length between } E_1 \text{ and } E_2.$$

Every allowed state is associated with two parameters: the wavevector k and the corresponding energy.

It is worthwhile examining the free-electron DOS in 1D space in order to obtain insight about the general formalism of deriving and interpreting the features of the electron DOS. We recall from Chapter 2 that the energy dispersion of the free-electron gas can be expressed as

$$E = \frac{\hbar^2 k^2}{2m} + E_o, \tag{5.1}$$

where m is the electron mass, $\hbar$ is the reduced Planck's constant, and E_o is the energy at the bottom of the parabolic band, which can be interpreted as the potential energy, and $E - E_o$ represents the electron kinetic energy. In a 1D solid, the number of states between E and $E + dE$ is the differential wave vector dk normalized to the length of one state:

$$g(E)\, dE\, L = 2 \times 2 \frac{dk}{2\pi/L}, \tag{5.2}$$

where L is the length of the 1D system and $2\pi/L$ is the length of one k-state; the factor of 4 in the numerator accounts for spin degeneracy and negative k-space.

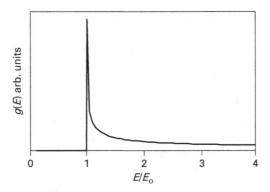

Fig. 5.1   The DOS of free electrons in a 1D solid.

Hence, it follows that a general formula for the electron DOS in a 1D solid is

$$g(E) = \frac{2}{\pi} \frac{dk}{dE}. \tag{5.3}$$

Applying the formula to Eq. (5.1) allows us to derive the DOS of free electrons in a 1D solid:

$$g(E) = \begin{cases} \dfrac{2}{h}\sqrt{\dfrac{2m}{E - E_o}} & E > E_o \\ 0 & E \leq E_o \end{cases}, \tag{5.4}$$

which is shown graphically in Figure 5.1.

There are several insights we gain visually from the free-electron DOS in 1D, including:

(i) There are no states inside the bandgap ($E < E_o$), a general truth applicable to all ideal crystalline solids.

(ii) There are no states at $E = E_o$, which, although it might not be obvious, can be deduced from fundamental postulates in quantum mechanics. In quantum mechanics, electrons are only allowed to exist in states that have a positive and finite kinetic energy ($E - E_o > 0$) leading to moving states. If states existed at $E = E_o$, the electrons in those states would be stationary,[1] implying that we can determine their position precisely, which would violate Heisenberg's uncertainty principle. Although the DOS diverges as $E \to E_o$, it is actually never infinite because $E = E_o$ is not an allowed energy. The divergence of the DOS is also known as a van Hove singularity (VHS), and hence E_o

---

[1] The velocity of an electron is formally related to the DOS in Section 5.6 where we find that $E = E_o$ corresponds to states with zero velocity for parabolic-like dispersions.

corresponds to the energy of the singularity. Van Hove singularities are key features of the electron DOS of 1D solids with a parabolic-like dispersion.

(iii) For $E > E_o$, the DOS $g(E) \sim 1/\sqrt{E - E_o}$, another key characteristic of 1D crystalline solids with parabolic-like dispersion.

In summary, there are two key features of the DOS which are particular to 1D solids in the limit of free electrons: VHS and the DOS inverse square-root dependence on energy. Interestingly, these two features are present in the DOS of electrons and holes in CNTs; as such, a CNT can be considered a 1D solid.

5.3 Density of states of zigzag nanotubes

The CNT electronic structure has many 1D subbands; as such, the total DOS g_{tot} at a given energy is the sum of the contributions from the DOS of each subband. Consider an infinitely long CNT for mathematical ease:[2] the total DOS and the DOS per subband are respectively

$$g_{\text{tot}}(E) = \sum_{j=1}^{N} g(E,j) \tag{5.5}$$

and

$$g(E,j) = \frac{1}{\pi} \left| \frac{\partial E}{\partial k} \right|^{-1} = \frac{1}{\pi} \left| \frac{\partial k}{\partial E} \right|, \tag{5.6}$$

where N is the number of subbands in the CNT band structure.

For simplicity, we shall first derive the DOS for zigzag $(n, 0)$ nanotubes, then armchair (n, n) nanotubes, and finally arbitrary chiral (n, m) nanotubes. The E–k dispersion of zigzag nanotubes is

$$E(k,j) = \pm \gamma \sqrt{1 + 4\cos\left(\frac{\sqrt{3}ka}{2}\right) \cos\left(\frac{\pi j}{n}\right) + 4\cos^2\left(\frac{\pi j}{n}\right)}, \tag{5.7}$$

where j is the subband integer index, which ranges from 1 to $2n$, a is the graphene Bravais lattice constant (~ 2.46 Å), and γ is the nearest neighbor overlap energy and nominally ~ 3.1 eV. The positive and negative prefixes refer to the band structures of the conduction (π^*) and valence (π) bands respectively. The intrinsic Fermi energy is 0 eV. To derive the DOS for zigzag CNTs g_{zz}, it is more convenient to

---

[2] Recall from Section 4.4 that an infinitely long CNT allows us to employ continuum mathematics for k-values in the axial direction, which remains a good approximation for nanotubes as short as 10 nm.

evaluate the RHS of Eq. (5.6), because $\partial k/\partial E$ results in an expression which is an explicit function of E. The wavevector for a zigzag CNT is

$$k = \pm \frac{2}{\sqrt{3}a} \cos^{-1} \left\{ \frac{1}{4} \sec \left(\frac{\pi j}{n} \right) \left[\frac{E^2}{\gamma^2} - 2\cos \left(\frac{2\pi j}{n} \right) - 3 \right] \right\}. \tag{5.8}$$

For deriving the DOS, it is easiest to differentiate the positive branch of the wavevector with respect to energy according to Eq. (5.6). It follows that, by organizing the resulting denominator, the DOS due to positive wavevectors can be written in the form given by

$$g_{zz}^{+}(E,j) = \frac{4}{\sqrt{3}a\pi} \frac{|E|}{\sqrt{(E^2 - E_{vh_1}^2)(E_{vh_2}^2 - E^2)}} \tag{5.9}$$

for $C_b \leq E \leq C_t$ for the conduction band and $V_b \leq E \leq V_t$ for the valence band. C_b and C_t are the bottom and top of the conduction band respectively, and V_b and V_t are the bottom and top of the valence band respectively. The notation C_{bi} and C_{ti} will refer to the bottom and top of the ith subband respectively, and vice versa for the valence band (see Figure 5.2).[3] E_{vh1} and E_{vh2} (the roots of the denominator) are the energies of the VHS in zigzag nanotubes and define the energy space where the DOS is real. Owing to mirror symmetry of the $E-k$ relationship, the DOS for the negative branch of the wavevector is identical to g_{zz}^{+}. Hence, the complete DOS from both branches of the wavevector for the jth subband is

$$g_{zz}(E,j) = \frac{4\alpha}{\sqrt{3}a\pi} \frac{|E|}{\sqrt{(E^2 - E_{vh1}^2)(E_{vh2}^2 - E^2)}}. \tag{5.10}$$

Here, α accounts for the Brillouin zone mirror symmetry or degeneracy.[4] Specifically, $\alpha = 1$ if E is energy at the Brillouin zone center (since the Γ-point center is common to both branches of the wavevector), otherwise $\alpha = 2$. The Brillouin zone is illustrated in Figure 5.2. Moreover, the distinction in α is only relevant to the first subband of metallic nanotubes because they possess moving states at the Γ-point. Practically speaking, for semiconducting zigzag CNTs (or subbands with parabolic-like curvature), α is always 2, since no states are allowed at the Γ-point because of the zero velocity, which violates the uncertainty principle.

---

[3] Electro-optical applications of CNTs mostly involve C_{bi} and V_{ti}, with little concern for C_{ti} and V_{bi} which are at higher energies.

[4] Keeping track of all the degeneracies or symmetries present when calculating the properties of nanotubes can quite easily tax the brain, and sometimes lead to a factor of 2 dilemma. The family of degeneracies includes spin degeneracy, subband degeneracy, point degeneracy at the Fermi energy, and Brillouin zone mirror symmetry (α). The key to becoming comfortable with degeneracies is to personally rederive the equilibrium properties such as the DOS from the band structure.

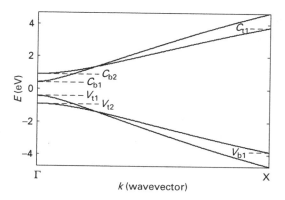

Fig. 5.2 Illustrative band structure of a nanotube displaying only the positive half of the Brillouin zone, and identifying the symmetry points Γ (center of BZ) and X (end of BZ). This particular case is for a (13, 0) CNT including only two subbands of the valence and conduction bands, and graphically indicating C_{bi}, C_{ti}, V_{ti}, and V_{bi}.

It is now timely to formally introduce the formulas of all the energy parameters defined so far in this section. The VHS are given by

$$E_{\text{vh1}}(j) = \pm\gamma \left|1 + 2\cos\left(\frac{j\pi}{n}\right)\right|, \tag{5.11}$$

$$E_{\text{vh2}}(j) = \pm\gamma \left|1 - 2\cos\left(\frac{j\pi}{n}\right)\right|. \tag{5.12}$$

We observe that the energies of the VHSs are chirality dependent and periodic with period $2n$, which is because there are identically $2n$ unique 1D subbands in the zigzag CNT Brillouin zone. E_{vh1} is located at the Γ-point and E_{vh2} is in the second Brillouin zone. These two singularities generally reflect the two minima/maxima present in a periodic sinusoid. In the case of a nanotube, it is the quasi-sinusoidal energy dispersion that leads to the VHS. The all-important bandgap is directly determined from E_{vh1} of the first subband. Let us call the subband index for the first subband j_1, then the bandgap is

$$E_{\text{g}} = 2C_{b1} = 2E_{\text{vh1}}(j_1) = 2\gamma \left|1 + 2\cos\left(\frac{\pi j_1}{n}\right)\right|. \tag{5.13}$$

The first subband index was determined in Chapter 4 to be $j_1 = \text{round}(2n/3)$, where round is a function that converts its argument to the nearest integer. A simpler approximation (useful to memorize) relating the bandgap and diameter explicitly was also derived in Chapter 4 to be

$$E_{\text{g}} = 2\gamma \frac{\pi}{\sqrt{3}n} = 2\gamma \frac{a_{\text{C–C}}}{d_{\text{t}}}, \tag{5.14}$$

where $a_{\text{C–C}}$ is the carbon–carbon bond length ($\sim$1.42 Å).

In order to deduce an equation for the bottom and top of the bands, another energy definition is required for completeness:

$$E_X(j) = \pm\gamma\sqrt{1 + 4\cos^2\left(\frac{\pi j}{n}\right)}.$$ (5.15)

E_X is the energy at the boundary of the first Brillouin zone (X-point in Figure 5.2), which is calculated by plugging $k = \pi/\sqrt{3}a$ into Eq. (5.7). Depending on the subband, the bottom or top of the subband is either at the Γ-point or at the X-point. Therefore:

$$C_b = \min(E_{vh1}, E_X) = -V_t,$$ (5.16)

$$C_t = \max(E_{vh1}, E_X) = -V_b,$$ (5.17)

where min( ) and max( ) are respectively the minimum and maximum of their arguments.

With the formalities of the energy definitions taken care of, we can now explore the working forms for the DOS for both metallic and semiconducting nanotubes. For metallic zigzag nanotubes, the DOS at the Fermi energy (located at $k = 0$) is of primary interest. The band index for the first subband is $j = 2n/3$ with a fourfold energy degeneracy at the Fermi energy (a factor of 2 due to subband degeneracy and another factor of 2 due to point degeneracy of the conduction and valence bands). Substituting $j = 2n/3$ and $k = 0$ into Eq. (5.10) gives the DOS at the Fermi energy (in short, the metallic DOS) for all metallic zigzag nanotubes:

$$g_0 = g_{zz}(E_F) = \frac{8}{\sqrt{3}a\pi\gamma} \sim 2 \times 10^9 \text{ m}^{-1}\text{ eV}^{-1} = 2 \text{ nm}^{-1}\text{ eV}^{-1}.$$ (5.18)

Notably, g_0 is independent of chirality and can be viewed as a material constant. Invariably, because it is independent of chirality, we might suspect that Eq. (5.18) will apply to *all* metallic nanotubes of arbitrary chirality, and this is in fact the case.[5]

Semiconducting zigzag CNTs can employ a simpler form of Eq. (5.10) for routine analysis. For the lowest subbands which are of greatest relevance, E_{vh2} is much larger than the applicable electron energies (Eq. (5.12) $> \gamma \gg E$). Hence, in the limit $E_{vh2}^2 \gg E^2$, Eq. (5.10) reduces to

$$g_{zz}(E,j) = \frac{8}{\sqrt{3}a\pi E_{vh2}}\frac{|E|}{\sqrt{E^2 - E_{vh1}^2}} = \frac{g_0\gamma}{E_{vh2}}\frac{|E|}{\sqrt{E^2 - E_{vh1}^2}}.$$ (5.19)

[5] R. Saito, G. Dresselhaus and M. S. Dresselhaus, *Physical Properties of Carbon Nanotubes* (Imperial College Press, 1998) Chapter 4.

Another approximation comes into play by studying the relationship between the two VHSs (Eq. (5.11) and Eq. (5.12)) for the lowest subbands, which after some algebra can be expressed as $E_{vh2} \approx E_{vh1} + 2\gamma$. 2γ is around 6 eV, which to first-order approximation can be considered much greater than common values of E_{vh1}. Therefore, the semiconducting zigzag DOS can be written in a simpler form useful as a working formula, especially for routine hand-analysis:

$$g_{zz}(E,j) \approx \frac{g_0}{2} \frac{|E|}{\sqrt{E^2 - E_{vh1}^2}}. \tag{5.20}$$

The DOS for semiconducting and metallic zigzag nanotubes (Eq. (5.10)) are shown in Figure 5.3. A noteworthy insight is that the square of the energy terms in the denominator of the expression for the DOS is due to the electron–hole symmetry present in the NNTB band structure of CNTs leading to mirror symmetry between the conduction and valence bands' DOS visually evident in Figure 5.3. Another insight is that the nanotube DOS generally reflects the DOS expected of free electrons in a 1D crystal. In the ideal 1D free-electron gas, there is a VHS and

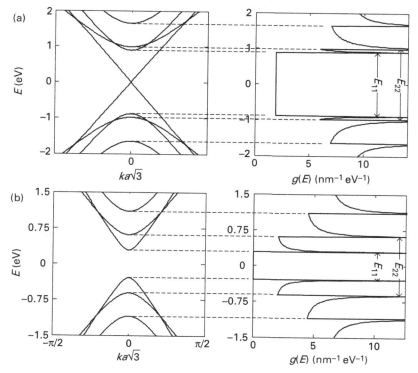

Fig. 5.3 The electronic DOS for selected (a) metallic and (b) semiconducting zigzag nanotubes. For semiconducting (metallic) nanotubes, the DOS around 0 eV is zero (non-zero). $E_{11}, E_{22}, \ldots, E_{nn}$ refers to the energies of optical transitions between the nth pair of VHSs.

the functional relation $g(E) \sim 1/\sqrt{E}$. In CNTs, we have several VHSs (arising from the several subbands) in both the conduction and valence bands, and the functional relation $g(E) \sim 1/\sqrt{E}$ beyond each VHS. Optically excited electron transitions between pairs of VHSs are an active area of experimental research, and often employed as a technique to identify the chirality of carbon nanotubes.[6]

It is worthwhile probing further on some important features of the zigzag DOS analytically to gain additional understanding of the electro-physical properties of CNTs. To that end, two auxiliary results are derived: the energy difference between the bottom of the first and second subbands and the energy bandwidth between the top and bottom of the first subband. The derivation focuses on the conduction band; however, the results apply equally to the valence band, owing to the electron–hole symmetry of the nanotube band structure. The energy separation between the first and second subbands is valuable to derive because it is of significance when determining the number of subbands that contribute to carrier density and transport in CNT. Let us label the subband index for the second subband j_2; thus, the bottom of the second subband can be expressed as

$$C_{b2} = E_{vh1}(j_2) = \gamma \left| 1 + 2\cos\left(\frac{\pi j_2}{n}\right) \right|. \tag{5.21}$$

The second subband index can be written in terms of the first subband index:

$$j_2 = j_1 - 1 \approx \left(\frac{2n}{3} + \frac{1}{3}\right) - 1 = \frac{2(n-1)}{3}, \tag{5.22}$$

where the factor $[(2n/3) + (1/3)]$ is the linear approximation for j_1 (Eq. (4.40)), which when substituted into Eq. (5.21) yields

$$C_{b2} \approx \gamma \left| 1 + 2\cos\left(\frac{2\pi}{3} - \frac{2\pi}{3n}\right) \right| = \gamma \left| 1 + 2\cos\left[\frac{2\pi}{3}\left(1 - \frac{1}{n}\right)\right] \right| \tag{5.23}$$

For relatively large n, $1/n$ is a small perturbation about unity; therefore, a first-order Taylor series expansion[7] of Eq. (5.23) centered at $2\pi/3$ in the cosine argument is sufficiently accurate, leading to the simple expression

$$C_{b2} = E_{vh1}(j_2) \approx \frac{2\pi\gamma}{\sqrt{3n}} = E_g. \tag{5.24}$$

Recalling that the bottom of the first subband is half of the bandgap, we can therefore conclude that the energy separation E_Δ between the bottoms of the first

[6] S. Bachilo, M. S. Strano, C. Kittrell, R. H. Hauge, R. E. Smalley and R. B. Wersman,
      Structure-assigned optical spectra of single-walled carbon nanotubes. *Science*, **298** (2002) 2361–6.
[7] The Taylor expansion is of $\gamma(1 + 2\cos[x])$ about $x = 2\pi/3$, where $x = (2\pi/3 - 2\pi/3n)$.

two subbands is

$$E_\Delta = C_{b2} - C_{b1} \approx \frac{E_g}{2}. \tag{5.25}$$

Lastly, we derive the relationship between the top and bottom of the first subband, conveniently referred to as the subband bandwidth E_{BW}. Formally.

$$E_{BW}(j_1) = E_X(j_1) - E_{vh1}(j_1). \tag{5.26}$$

By means of Eq. (5.11) and Eq. (5.15) for the bottom and top of the first subband respectively, and employing $j_1 \sim (2n/3) + (1/3)$ for the subband index, and a Taylor series expansion for Eq. (5.26) as exploited earlier for Eq. (5.23), the first subband bandwidth can be expressed as

$$E_{BW}(j_1) \approx \sqrt{2}\gamma \left(1 + \frac{\pi}{2\sqrt{3}n} - \frac{\pi}{\sqrt{6}n}\right) \approx \sqrt{2}\gamma = 1.4\gamma, \tag{5.27}$$

which is a useful approximation for zigzag nanotubes with chiral index n as small as 13 (diameter $\sim$1 nm).

5.4 Density of states of armchair nanotubes

Following the same approach used for the derivation of the zigzag nanotube DOS, the DOS for the jth subband for an armchair nanotube can be computed from its energy dispersion E given by

$$E(k,j) = \pm\gamma\sqrt{1 + 4\cos\left(\frac{\pi j}{n}\right)\cos\left(\frac{ka}{2}\right) + 4\cos^2\left(\frac{ka}{2}\right)}, \tag{5.28}$$

which can be rewritten for the wavevector:

$$k = \pm\frac{2}{a}\cos^{-1}\left[-\frac{1}{2}\cos\left(\frac{\pi j}{n}\right) \pm \frac{1}{2}\sqrt{\frac{E^2}{\gamma^2} + \cos^2\left(\frac{\pi j}{n}\right) - 1}\right]. \tag{5.29}$$

It follows that evaluating Eq. (5.6) results in the electronic DOS for armchair CNTs.

$$g_{ac}(E,j) = \frac{8}{a\pi} \frac{|E|}{\sqrt{E^2 - E_{vh1}^2}\left(-A_1 + \sqrt{E^2 - E_{vh1}^2}\right)\left(A_2 - \sqrt{E^2 - E_{vh1}^2}\right)} \tag{5.30}$$

for $C_b \leq E \leq C_t$ for the conduction band and $V_b \leq E \leq V_t$ for the valence band. E_{vh1} is an armchair VHS:

$$E_{vh1}(j) = \pm\gamma \left| \sin\left(\frac{\pi j}{n}\right) \right|, \tag{5.31}$$

$$E_{vh2}(j) = \pm\gamma \sqrt{5 - 4\cos\left(\frac{\pi j}{n}\right)}, \tag{5.32}$$

where E_{vh2} is another armchair nanotube VHS that is not explicit in Eq. (5.30). The energy parameters A_1 and A_2 have been defined to make Eq. (5.30) tractable:

$$A_1(j) = \gamma \left[-2 + \cos\left(\frac{\pi j}{n}\right) \right], \tag{5.33}$$

$$A_2(j) = \gamma \left[2 + \cos\left(\frac{\pi j}{n}\right) \right]. \tag{5.34}$$

It is necessary to calculate the energy at Brillouin zone boundaries to determine the bottom and top of the subbands. The Γ-point energy is computed by setting $k = 0$ in Eq. (5.28), and the X-point energy is computed by setting $k = \pi/a$:

$$E_\Gamma(j) = \pm\gamma \sqrt{5 + 4\cos\left(\frac{\pi j}{n}\right)}, \tag{5.35}$$

$$E_X(j) = \pm\gamma. \tag{5.36}$$

For armchair CNTs, the bottom of the conduction subband could be either within the Brillouin zone or at the X-point, depending on the subband index, and the top of the subband is the larger of the Γ-point and X-point energies:

$$C_t = \max(\gamma, E_\Gamma) = -V_b \tag{5.37}$$

$$C_b = \begin{cases} E_{vh1} = -V_t & n - \text{floor}(n/2) \leq j \leq n + \text{floor}(n/2) \\ \gamma = -Vt & n - \text{floor}(n/2) > j > n + \text{floor}(n/2) \end{cases}, \tag{5.38}$$

where the floor( ) function rounds its argument to the lowest integer.

Since all armchair CNTs are metallic (even with nanotube curvature effects included), the DOS at the Fermi energy is of great significance. The index for the lowest subband is $j = n$ located at $k = \pm 2\pi/3a$. The calculated Fermi DOS is

$$g_{ac}(E_F) = g_0 = \frac{8}{\sqrt{3}a\pi\gamma} \sim 2 \times 10^9 \text{ m}^{-1} \text{ eV}^{-1} = 2 \text{ nm}^{-1} \text{ eV}^{-1}, \tag{5.39}$$

which is identical to the metallic zigzag DOS, further confirming the general conclusion that all metallic nanotubes have the same Fermi DOS. Figure 5.4 shows the DOS of selected armchair CNTs.

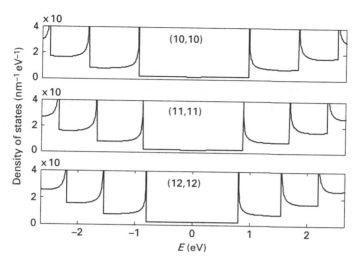

Fig. 5.4   The electronic DOS for selected armchair nanotubes. All armchair nanotubes are metallic with a finite DOS at the Fermi energy (0 eV).

5.5 Density of states of chiral nanotubes and universal density of states for semiconducting CNTs

The DOS of arbitrary chiral nanotubes can be determined from the same basic formula (Eq. (5.6)), with the dispersion of chiral CNTs given by

$$E(k)^{\pm} = \pm \gamma \sqrt{1 + 4\cos\left(\frac{\sqrt{3}a}{2}k_x\right)\cos\left(\frac{a}{2}k_y\right) + 4\cos^2\left(\frac{a}{2}k_y\right)}, \qquad (5.40)$$

where

$$k_x = \frac{2\pi\sqrt{3}aj(n+m)C_h + a^3k(n^3 - m^3)}{2C_h^3},$$

$$k_y = \frac{\sqrt{3}ak(n+m)C_h + 2\pi aj(n-m)}{2C_h^2}, \qquad (5.41)$$

and C_h is the magnitude of the chiral vector. The solution for the DOS requires numerical techniques, because $g(E)$ for chiral CNTs cannot be solved explicitly as a function of energy in an algebraic manner. Nonetheless, the insights gained from the study of the DOS for zigzag and armchair nanotubes can be extended to chiral CNTs. For example, the DOS for the metallic chiral nanotube at the Fermi energy is given by g_0 (Eq. (5.18)), because the gradient of the dispersion at the Fermi energy is a constant property of the linear dispersion of graphene (also called the Dirac cone), which is independent of chirality. Additionally, VHS exist at the bottom and top of subbands that possess curvature. Moreover, between VHSs, the DOS will exhibit the inverse square-root dependence $g(E) \sim 1/\sqrt{E}$.

Although the DOS for chiral nanotubes can be solved numerically, there exists a need for a simple analytical expression that would benefit essential applications, such as compact modeling of CNT devices and CNT sensor and circuit design. To that end, we might be able to extend the basic insight that the bandgaps of CNTs are primarily diameter dependent with negligible chirality dependence for this cause. This was demonstrated (in Chapter 4) by employing a semiconducting zigzag nanotube as a model CNT, deriving its bandgap and expressing it in terms of diameter, which revealed an $E_g \sim 1/d_t$ relation that applies equally well to arbitrary (n, m) chiral indices for $(n, m) > (7, 0)$. Similarly, to obtain a simple analytical expression for the DOS of chiral CNTs, a logical idea is to start with the analytical DOS for zigzag nanotubes, rewrite it as a function of diameter, and see how well it describes the numerically computed DOS for chiral nanotubes of similar diameters. In essence, we desire to replace the chirality dependence with diameter dependence for the DOS for zigzag CNTs and hope that the diameter-dependent DOS will be accurate for describing the DOS for arbitrary diameter (or arbitrary chirality) nanotubes. Before investing the effort to develop an analytical DOS for chiral CNTs, we can actually evaluate if our main idea will work. The focus here will be on semiconducting nanotubes, because it has previously been discussed that metallic nanotubes have a constant DOS independent of chirality and diameter at the Fermi energy, which is the energy of main interest for metallic or interconnect applications. Table 5.1 shows selected zigzag CNTs with diameters that span the practical range from ~1 to ~3 nm, and chiral CNTs with the closest comparable diameters to the zigzag nanotubes.

A comparison between the DOSs of the selected zigzag nanotubes and chiral nanotubes of similar diameters is presented in Figure 5.5, showing a strong likeness up until approximately the bottom of the third subband. This implies that the DOS for semiconducting zigzag CNTs can be used as a basis for developing a simple analytical expression for the DOS for chiral nanotubes of comparable diameters for the most important (lowest) subbands. The zigzag DOS (Eq. (5.19)) is

$$g_s(E, j) \approx \frac{g_0 \gamma}{E_{vh2}} \frac{|E|}{\sqrt{E^2 - E_{vh1}^2}}, \tag{5.42}$$

which we will now refer to as a diameter-dependent $g_s(E)$ to symbolize a *universal DOS*, an idea that was first espoused and discussed in the literature by Mintmire and White.[8] The energies of the VHS can be expressed explicitly in terms of the diameter by employing Eq. (5.11) and Eq. (5.12) and the $d_t = an/\pi$ relation for

[8] J. W. Mintmire and C. T. White, Universal density of states for carbon nanotubes. *Phys. Rev. Lett.*, **81** (1998) 2506–9.

Table 5.1. List of four groups of zigzag and chiral CNTs of comparable diameters

(n,m)	d_t(Å)	$(C_h/a)^2$	(n,m)	d_t(Å)	$(C_h/a)^2$
(13, 0)	10.18	169	(26, 0)	20.36	676
(8, 7)	10.18	169	(16, 14)	20.36	676
(12, 2)	10.27	172	(21, 8)	20.31	673
(19, 0)	14.88	361	(38, 0)	29.76	1444
(16, 5)	14.88	361	(32, 10)	29.76	1444
(18, 2)	14.94	364	(37, 2)	29.79	1447

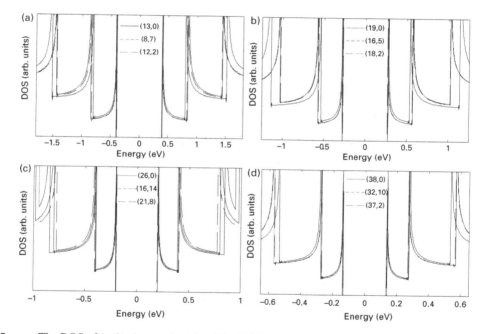

Fig. 5.5   The DOS of $(n,0)$ zigzag nanotubes (Eq. (5.10)) compared with numerically computed DOS of (n,m) chiral CNTs of similar diameters showing a strong likeness up until approximately the bottom of the third subband.

zigzag CNTs (see Table 4.1):

$$E_{vh1}(j) = \pm\gamma \left| 1 + 2\cos\left(\frac{ja}{d_t}\right) \right|; \tag{5.43}$$

$$E_{vh2}(j) = \pm\gamma \left| 1 - 2\cos\left(\frac{ja}{d_t}\right) \right|. \tag{5.44}$$

In summary, Eqs (5.42)–(5.44) constitute *universal DOS formulas* for semiconducting nanotubes of arbitrary diameters or chiralities with good accuracy up to the bottom of the third subband, i.e. $E < C_{b3}$ for the conduction band and $E > V_{t3}$ for the valence band. A desirable outcome of the universal DOS is that the properties of CNT that are directly related to the DOS or gradient of the dispersion would be primarily diameter dependent with negligible chirality dependence. Another way to articulate this desirable outcome is that high-symmetry (or simpler) zigzag CNTs can be used as a model nanotube in order to gain insight and understanding and the results obtained can be generally applied to arbitrary chiral nanotubes of comparable diameters. For example, the (19, 0) CNT, which has a diameter of about 1.5 nm, is often used as a model nanotube in the literature for nanotube transistor studies.

5.6   Group velocity

The group velocity of an electron in essence informs us of the speed of the electron, and it is a useful parameter that aids us in determining how fast energy or information is propagating in the ideal case when electrons do not suffer any scattering that might impede their travels. It is often closely associated with the maximum information (or signal) frequency that the electrons can transmit. Formally, the electron group velocity is defined as

$$|v(E,j)| = \frac{1}{\hbar}\left|\frac{\partial E}{\partial k}\right| = \frac{1}{\hbar}\left|\frac{\partial k}{\partial E}\right|^{-1}. \tag{5.45}$$

The velocity is a 1D vector quantity; hence, care has to be exercised to keep track of both its magnitude and direction. We learned from Chapter 4 that the band structure of CNTs possesses horizontal mirror symmetry (also called Brillouin zone symmetry) and vertical mirror symmetry (also called electron–hole symmetry). The electron–hole symmetry allows us the flexibility to analyze the electron velocity and apply the results equally to holes. The Brillouin zone mirror symmetry entails that the branches of the dispersion in the negative half of the Brillouin zone (known as negative branches) have the negative of the slope of the positive branches. That is, for every allowed state with an energy E in the positive branch and a velocity v, there exists a state in the negative branch also with an energy E but with a velocity $-v$. In the literature, when describing transport in CNT devices, carriers with a positive velocity are sometimes referred to as forward or right-moving carriers, while carriers with a negative velocity are labeled backwards or left-moving carriers.

Of frequent utility is the group velocity at the Fermi energy in metallic nanotubes, which by solid-state physics convention is referred to as the Fermi velocity v_F and can be determined by evaluating the velocity for metallic zigzag nanotubes at the Fermi energy ($j = 2n/3, k = 0$).[9] The first electron subband energy dispersion for

---

[9] A metallic zigzag CNT is chosen for simplicity. The results for the Fermi velocity apply in general to *all* metallic nanotubes.

a metallic zigzag CNT is

$$E(k) = \sqrt{2}\gamma \sqrt{1 - \cos\left(\frac{\sqrt{3}ka}{2}\right)}. \tag{5.46}$$

It follows that the Fermi velocity is

$$v_F = \frac{1}{\hbar}\frac{\partial E}{\partial k}\Big|_{k=0} = \frac{\sqrt{3}a\gamma}{4\hbar}\frac{\sin(\sqrt{3}ka/2)}{\sin(\sqrt{3}ka/4)}, \tag{5.47}$$

for which, in the limit $k \to 0$, $\sin(x)$ becomes x and the magnitude of the Fermi velocity simplifies to

$$|v_F| = \frac{\sqrt{3}a\gamma}{2\hbar}. \tag{5.48}$$

With $\gamma \sim 3.1$ eV, the velocity is $v_F \sim 10^6$ m s^{-1} which is, not surprisingly, identical to v_F for graphene, since the gradient of the dispersion at the Fermi energy is the same for both graphene and nanotubes. Compared with bulk metals, the CNT v_F is within a factor of two of most conventional metals.[10]

For semiconducting zigzag nanotubes, the magnitude of the group velocity per subband can be expressed in terms of the (Eq. (5.10)):

$$|v(E,j)| = \frac{1}{\hbar}\left(\frac{\partial k}{\partial E}\right)^{-1} = \frac{\alpha}{\hbar\pi g_{zz}(E,j)}. \tag{5.49}$$

α is necessary in the numerator to normalize out the Brillouin zone mirror symmetry or degeneracy. For a *universal* semiconducting nanotube, as discussed in Section 5.5, g_{zz} can be replaced with g_s, and $\alpha = 2$. Note that the velocity at the bottom and top of quasi-parabolic subbands (subbands with a curvature), such as the lowest subbands, vanishes to zero. We can conclude that, owing to Heisenberg's uncertainty principle, which requires electron waves to have a finite speed, such states (with zero velocity) are not legitimate states. They are present as an artifact in the mathematical formalism only because we have employed continuum mathematics to describe what is, in reality, a discrete set of states.

5.7   Effective mass

The electron or hole effective mass m^* is a widely used concept in semiconductor physics and has been shown to be useful in modeling and studying low-energy

---

[10] The Fermi velocity of bulk metals is conveniently tabulated in N. W. Ashcroft and N. D. Mermin, *Solid State Physics* (chapter 2) (Brooks/Cole, 1976). For example, v_F for aluminum, copper, and gold are 2×10^6 m s^{-1}, 1.6×10^6 m s^{-1}, 1.4×10^6 m s^{-1} respectively.

electrical transport in CNTs.[11] Instead of employing the conventional formula $m^* = \hbar^2/(\partial^2 E/\partial k^2)$, which is an explicit function of k, we derive it explicitly as a function of energy and use the previous results for the DOS. Recalling the wave–particle duality in quantum mechanics, the effective mass is derived by equating the particle Newtonian force to the wave force:

$$F = \frac{\partial p}{\partial t} \Rightarrow \hbar \frac{\partial k}{\partial t} = m^* \frac{\partial v}{\partial t}, \tag{5.50}$$

where F is the force, p is the momentum, and t is time. This leads to an expression for m^* in terms of the dispersion:

$$m^*(E,j) = \hbar \frac{\partial k}{\partial v} = \hbar \frac{\partial k}{\partial E}\frac{\partial E}{\partial v} = \hbar \frac{\partial k}{\partial E}\left[\frac{1}{\hbar}\frac{\partial}{\partial E}\left(\frac{\partial E}{\partial k}\right)\right]^{-1}. \tag{5.51}$$

Substituting the DOS for the energy gradients, a general expression for m^* for a 1D subband is

$$m^*(E,j) = \hbar^2 \pi^2 g(E,j)\left[\frac{\partial}{\partial E}\left(\frac{1}{g(E,j)}\right)\right]^{-1}. \tag{5.52}$$

An elementary expression is achievable for semiconducting zigzag nanotubes by substituting Eq. (5.10) and performing the necessary differentiation and simplifications, including normalizing out the Brillouin zone mirror degeneracy; the semiconducting nanotube effective mass is

$$m^*(E,j) = \frac{16\hbar^2}{3a^2}\frac{E^3}{E_{vh1}^2 E_{vh2}^2 - E^4}. \tag{5.53}$$

Employing the relation $E_{vh1} = E_g/2$ and the useful approximation $E_{vh2} \approx E_{vh1} + 2\gamma$, the effective mass for the first electron subband (with $E > E_g/2$) is

$$m^*(E,j) \approx \frac{256\hbar^2}{3a^2}\frac{E^3}{E_g^2(E_g + 4\gamma)^2 - 16E^4}, \tag{5.54}$$

which is certainly not a constant, as is the case for the low energies around the bottom (top) of the conduction (valence) band in bulk semiconductors. Nonetheless, the effective mass at energies approaching the band minimum, which is

$$m^*\left(\frac{E_g}{2}\right) = \frac{4\hbar^2}{3\gamma a^2}\frac{E_g}{2\gamma + E_g}, \tag{5.55}$$

[11] S. O. Koswatta, N. Neophytou, D. Kienle, G. Fiori and M. S. Lundstrom, Dependence of dc characteristics of CNT MOSFETs on bandstructure models. *IEEE Trans. Nanotechnol.*, **5** (2006) 368–72.

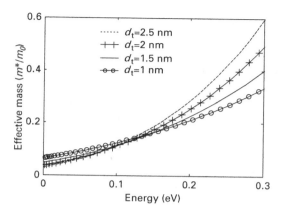

Fig. 5.6 Effective mass of the first conduction subband for 1 nm (13, 0), 1.5 nm (19, 0), 2 nm (26, 0), and 2.5 nm (32, 0) nanotubes. The energies have been normalized such that the band minimum is at 0 eV.

has been used to model transport in CNT field-effect transistors.[11] The band minimum m^* has a sublinear dependence on bandgap, and for a 1.5 nm (19, 0) CNT has a value $m^* \sim 0.046m_0$, ($\gamma = 3.1$ eV, $E_g \sim 0.58$ eV, and m_0 is the rest mass). A plot of the effective mass for selected semiconducting nanotubes is shown in Figure 5.6.

5.8 Carrier density

The density of charge carriers (electrons or holes) is a central property of semiconductors and is often invaluable in determining charge and heat transport in nanotube devices. We will focus on the electron carrier density with the implicit understanding that the mathematical techniques and results apply equally well to holes owing to the electron–hole symmetry in the CNT band structure. The zigzag semiconducting nanotube is most amenable to analytical treatment and, as such, its DOS will be in use in the subsequent analysis. However, the results for the carrier density can be applied to arbitrary chiral nanotubes, as discussed at the end of this section. The electron carrier density n_e per subband is the total number of occupied states in the subband:

$$n_e(j) = \int_{C_b}^{C_t} F(E)g_{zz}(E,j)\, dE. \tag{5.56}$$

For the lowest subbands, C_b and C_t are equal to E_{vh1} and E_X respectively. $F(E)$ is the Fermi–Dirac distribution, given by

$$F(E) = \frac{1}{1 + e^{(E-E_F)/kT}}, \tag{5.57}$$

where E_F is the Fermi energy,[12] k is Boltzmann's constant,[13] and T is the temperature. Equation (5.56) does not have an exact analytical solution. However, under certain approximations, an algebraic solution is accessible. To begin, we make the approximation that the top of the band (C_t) is much greater than the bottom of the band (C_b) and, owing to the rapidly decaying tail of the Fermi function, there is negligible error taking the upper limit of the integral to infinity. This approximation is justified by inspection of Eq. (5.27). Using Eq. (5.19) for g_{zz}, the simplified the carrier density integral is

$$n_e(j) = \frac{g_0\gamma}{E_{vh2}} \int_{E_{vh1}}^{\infty} \frac{1}{1 + e^{(E-E_F)/kT}} \frac{E}{\sqrt{E^2 - E_{vh1}^2}} dE,$$

$$n_e(j) = \frac{g_0\gamma E_{vh1}}{E_{vh2}} \int_1^{\infty} \frac{1}{1 + \exp(zt - z_F)} \frac{t}{\sqrt{t^2 - 1}} dt, \qquad (5.58)$$

where we have made some substitutions for simplicity ($t = E/E_{vh1}$, $z = E_{vh1}/kT$, and $z_F = E_F/kT$).

Alas, this simplified integral does not have a closed-form analytical solution. To proceed, we will develop formulas for the carrier density based on restricting the position of the Fermi energy relative to the band bottom. Perhaps the most elementary restriction of E_F is the non-degenerate assumption which restricts the Fermi energy to $3kT$ below the band bottom ($E_F \leq C_b - 3kT$) and is broadly used in bulk semiconductor physics.[14] Under this condition, the Fermi–Dirac distribution can be approximated by the Maxwell–Boltzmann distribution $F(E) \approx \exp[(E_F - E)/kT]$.[14] Invariably, owing to the rapidly decaying exponential tail of the Maxwell–Boltzmann distribution, only the first subband contributes an appreciable number of carriers. A graphical illustration of the DOS, the Fermi–Dirac distributions, and the implications of the non-degenerate approximation is shown in Figure 5.7. The resulting non-degenerate carrier density n_{e_nd} integral for the first subband is

$$n_{e\_nd} = \frac{2g_0\gamma}{E_{vh2}} \left(\frac{E_g}{2}\right) e^{E_F/kT} \int_1^{\infty} e^{-zt} \frac{t}{\sqrt{t^2 - 1}} dt, \qquad (5.59)$$

where the additional factor of 2 in the numerator is to account for the subband degeneracy and E_{vh1} has been substituted with $E_g/2$. The integrand has the form

---

[12] The Fermi energy is often used interchangeably with the phrase chemical potential μ in semiconductor literature. Formally, E_F is only defined at $T = 0$ K and μ is the technical jargon at finite temperatures. For convenience, we use E_F for all temperatures.

[13] The reader should be aware to distinguish the symbol k depending on the context, i.e. whether it is used to symbolize wavevector or Boltzmann's constant. It is often obvious when it is used for the latter, as it will be multiplied with temperature.

[14] R. F. Pierret, *Semiconductor Device Fundamentals* (Addison-Wesley, 1996), Chapter 2.

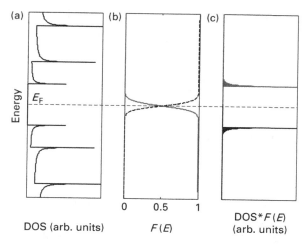

(a)   (b)   (c)

Energy

E_F

0   0.5   1

DOS (arb. units) $F(E)$ DOS*$F(E)$
                                      (arb. units)

Fig. 5.7   Illustrative plot of the DOS and Fermi–Dirac distribution at equilibrium. (a) Conduction
and valence band DOS. (b) $F(E)$ for electrons (solid) and holes (dashed). (c) DOS
multiplied by the Fermi–Dirac function (has the form of a modified Bessel function of the
second kind), showing that, essentially, only the first subband contributes appreciably to
carrier density at equilibrium.

of the modified Bessel function of the second kind, which looks similar to the
curve illustrated in Figure 5.7c. Fortunately, the integral has a solution given by[15]

$$n_{e\_nd} = \frac{g_0 \gamma E_g}{E_{vh2}} e^{E_F/kT} K_1 \left(\frac{E_g}{2kT} \right), \qquad (5.60)$$

where $K_1(z)$ is the modified Bessel function of the second kind of order one. An
accurate closed-form approximation for $K_1(z)$ is given as[16]

$$K_1(z) \approx \sqrt{\frac{\pi}{2z^3}} \frac{(1+2z)}{2} e^{-z}. \qquad (5.61)$$

It follows that the non-degenerate carrier density can be expressed algebraically as

$$n_{e\_nd} = n_i e^{E_F/kT}, \qquad (5.62)$$

$$n_i = \frac{g_0 \gamma (E_g + kT)}{E_{vh2}} \sqrt{\frac{\pi kT}{E_g}} e^{-E_g/2kT}, \qquad (5.63)$$

[15] The integral is interestingly the definition of the modified Bessel function of the second kind of
order one. See M. Abramowitz and I. A. Stegun, *Handbook of Mathematical Functions with
Formulas, Graphs and Mathematical Tables*, 10th edn (Washington, DC: US Government
Printing Office, 1972).

[16] D. Akinwande, Y. Nishi and H.-S. P. Wong, An analytical derivation of the density of states,
effective mass, and carrier density for achiral carbon nanotubes. *IEEE Trans. Electron Devices*, **55**
(2008) 287–97.

where n_i is the temperature-dependent intrinsic (or equilibrium) carrier density defined for $E_F = 0$. A more convenient form for hand analysis for n_i is achievable by substituting $E_{vh2} \approx E_{vh1} + 2\gamma$:

$$n_i = \frac{2g_0\gamma(E_g + kT)}{E_g + 4\gamma}\sqrt{\frac{\pi kT}{E_g}}e^{-E_g/2kT}. \qquad (5.64)$$

Likewise, the non-degenerate hole carrier density is

$$n_{h\_nd} = n_i\,e^{-E_F/kT}. \qquad (5.65)$$

Figure 5.8a shows the non-degenerate electron carrier density for semiconducting zigzag nanotubes where the exponential dependence seen in 3D semiconductors is also observed in 1D CNTs. Moreover, the formulas for the non-degenerate carrier density for CNTs (Eqs (5.62)–(5.65)) have essentially the same functional form as for 3D bulk semiconductors.[14] Quite interestingly, the form of the carrier density and its dependence on bandgap, temperature, and Fermi energy can be deduced simply by dimensional scaling of the 3D bulk carrier density, specifically by scaling the *partition function* in statistical mechanics from three dimensions to one dimension, as shown in the appendix of the article referenced in footnote 16.

A question of interest is how the intrinsic carrier density of CNTs compares with the intrinsic carrier densities of bulk semiconductors. To make this comparison possible, the CNT carrier density is normalized to the volume of the nanotube and compared against bulk semiconductors in Figure 5.8b. It is clear from the figure that nanotubes offer more carriers per unit volume, and this partly explains

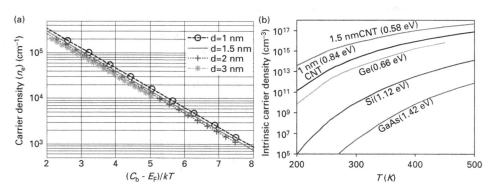

Fig. 5.8 (a) CNT carrier density using Eq. (5.60) as a function of the Fermi level at room temperature. Within the non-degenerate range ($C_b - E_F > 3kT$), the maximum |error| < 3.5 %, and |error| < 6% if Eq. (5.62) is used. (b) The intrinsic carrier density of ~1 nm (13, 0) and ~1.5 nm (19, 0) nanotubes (normalized to the volume of the nanotube), compared with the intrinsic carrier densities of bulk semiconductors.[14] The numbers in parentheses are the respective bandgaps.

the higher (dimensionally) normalized current in nanotube devices compared with conventional semiconductor devices.

The non-degenerate carrier density is a first step towards our initial goal of obtaining an analytical expression for the carrier density in CNTs. We now seek a more general expression for the carrier density that accounts for positions of the Fermi energy relative to the conduction band minima. To that end, Guo et al.[17] have shown that up to two subbands are significant in determining charge transport in practical nanotube transistors. This implies that, for the most part, only the first two subbands participate in contributing carriers to the overall carrier density. In that light, Liang et al.[18] developed a semi-empirical analytical formula for the carrier density in CNTs including the first two subbands. For convenience, we will refer to it as the two-subband carrier density, which is given as

$$
n_e \approx \sum_{j=1}^{2} n_e(j) = \sum_{j=1}^{2} n_i \, e^{E_g/2kT} \frac{e^{x_n}}{1 + A \, e^{\alpha x_n + \beta x_n^2}}, \quad x_n = \frac{2E_F - E_g}{2kT}, \quad (5.66)
$$

$$
n_h \approx \sum_{j=1}^{2} n_h(j) = \sum_{j=1}^{2} n_i \, e^{E_g/2kT} \frac{e^{x_p}}{1 + A \, e^{\alpha x_p + \beta x_p^2}}, \quad x_p = \frac{-E_g - 2E_F}{2kT}, \quad (5.67)
$$

where α, β, and A are fitting parameters and the sum is over the first two subbands.[19] Within the two-subband approximation the relative positions of the Fermi energy considered are $\max(E_F \leq C_{b2})$ for electrons and $\min(E_F \geq V_{t2})$ for holes (see Figure 5.2 for visual positions of C_{b2} and V_{t2} in the CNT band structure). For semiconducting zigzag nanotubes of diameters from 1 to 4 nm at temperatures within the practical range of 220 K to 375 K, the values of the fitting parameters with highest accuracy are $\alpha = 0.88$, $\beta = 2.41 \times 10^{-3}$, and $A = 0.63$, resulting in an error <6% compared with NNTB numerical computations. Fitting parameters for a greater range of CNT diameters and temperatures are also discussed in the work of Liang et al.[18] Figure 5.9 is a comparison between the numerical and analytical electron carrier density at room temperature for selected nanotubes showing strong agreement. An attractive property of the semi-empirical carrier density formulas is that they reduce to the non-degenerate carrier density formula when E_F is within the non-degenerate limit because the denominator of Eq. (5.66) approaches unity.

It is worthwhile keeping in mind that the non-degenerate and two-subband carrier density expressions can be applied to determine the carrier densities of arbitrary (n, m) semiconducting nanotubes because the energy range of interest

---

[17] J. Guo, A. Javey, H. Dai and M. Lundstrom, Performance analysis and design optimization of near ballistic carbon nanotube field-effect transistors. *IEEE IEDM Tech. Digest*, (2004) 703–6.

[18] J. Liang, D. Akinwande and H.-S. P. Wong, Carrier density and quantum capacitance for semiconducting carbon nanotubes. *J. Appl. Phys.*, **104** (2008) 064515.

[19] The reader should be aware that $j = 1$ or 2 in this context refers to the first or second subband, not the value of the subband index itself. For example, the value of the subband index for the first subband is round $(2n/3)$, where n is the zigzag chiral integer.

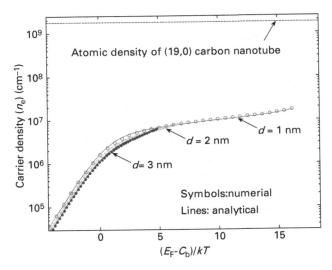

Fig. 5.9   Comparison of the two-subband analytical electron carrier density and numerical computation of the carrier density at room temperature showing good agreement (error $< \pm 5\%$). The dashed line is the atomic density of (19, 0) zigzag CNT for reference.

considered in the carrier density derivation clearly falls below the bottom of the third subband where the universal DOS idea applies.

5.9   Summary

We have explored the essential equilibrium properties of CNTs in this chapter. Much of the effort was to obtain an understanding of the gradient and curvature of the CNT band structure. For example, the band gradient is captured in the DOS, and the band curvature informs us of the effective mass of charge carriers. We found that the DOS which reflects the 1D nature of nanotubes is a very important property that is directly related to many other electronic properties, such as electron velocity and carrier density. It is not an exaggeration to assert that understanding the CNT DOS is essential to gaining insight about electron behavior in nanotube devices.

The key result of this chapter is the development of analytical equations to describe the equilibrium properties. Particularly useful is that, by employing the universal DOS idea, the equations or formulas developed can be applied to arbitrary (n, m) nanotubes for practical device applications. Talking about practical applications, it is important to keep in mind that interesting devices generally do not operate in equilibrium, because of external applied fields, which serves to provide us a degree of freedom to control electron transport. Nonetheless, the departure from equilibrium is often considered negligible or at least weak enough to be ignored to first order. This is especially true for low-energy excitation of electrons in nanotubes. In this limit, which is frequently referred to as quasi-equilibrium, the Fermi–Dirac function remains an accurate distribution of occupied states. At high energies, self-consistent numerical computation is typically employed to determine the non-equilibrium properties, such as the carrier velocity and carrier density.[20] Owing to scaling of conventional semiconductor technology, it is expected that the majority of device applications of CNTs will be at low operating energies.

5.10   Problem set

All the problems are intended to exercise and refine analytical techniques while providing important insights. If a particular problem is not clear, the reader is encouraged to re-study the appropriate sections. Invariably, some problems will involve making reasonable approximations or assumptions beyond what is already stated in the specific question in order to obtain a final answer. The reader should not consider this frustrating, because this is obviously how problems are solved in the real world.

5.1. Free-electron density of states in two and three dimensions.
Analysis and insights regarding the free-electron density of states in 1D space become more relevant and useful when compared with the characteristic density of states in 2D and 3D space. As a result, this exercise focuses on deriving the DOS for free electrons in higher dimensions.

(a) Derive the 2D DOS of free electrons.
(b) Derive the 3D DOS of free electrons.
(c) Sketch the 1D, 2D, and 3D DOS on the same graph; that is, $g(E)$ in arbitrary units versus energy.

5.2. Higher subbands in metallic CNTs.
A linear dispersion is often assumed for armchair nanotubes (representative of metallic CNTs) because at low energies only the first subband contributes to electron transport. To gauge the accuracy of this assumption, it is worthwhile determining the energy minimum of the second subband. This energy minimum can be viewed as a threshold above which the linear dispersion no longer holds.

(a) Determine analytically an expression for the second subband energy for armchair CNTs.
(b) What is the diameter dependence of the second subband energy?

---

[20] For an accessible discussion of non-equilibrium techniques for electron devices, see M. Lundstrom and J. Guo, *Nanoscale Transistors: Device Physics, Modeling and Simulation* (Springer, 2006).

(c) Compare the diameter dependence of the second subband energy with the diameter dependence of the bandgap of semiconducting CNTs. Is the functional dependence identical in both cases?

5.3. Density of states for armchair CNTs.

(a) Derive the DOS for armchair CNTs (representative of metallic CNTs) considering only the first subband.

(b) What is the effective mass for electrons in the first subband of an armchair nanotube?

(b) Derive an expression for the electron carrier density considering only the contributions from the first subband.

5.4. Higher subbands in semiconducting carbon nanotubes

For transistor applications, the first two subbands of a semiconducting CNT are often assumed to provide the lion's share of electrical current, particularly at low energies. In other words, the effects of higher subbands are typically neglected. The validity of this assumption strictly depends on the position of the Fermi level relative to the energy minima of higher subbands. For this reason, it is useful to have equations for the energy minima of the higher subbands, especially for small-bandgap CNTs.

(a) Determine analytically an expression for the third subband energy (in terms of the diameter) for a semiconducting zigzag CNT.

(b) If any approximations are made to arrive at an analytical expression, quantify the error compared with numerical evaluation of the third subband energy.

(c) What is the relation between the third subband energy and the bandgap?

5.5. Kataura plot for E_{ii}.

A plot of the transition energies between pairs of VHSs (E_{ii}) is commonly called a Kataura plot, after Hiromichi Kataura, who investigated these transition energies. A plot of these transition energies is often used to aid in the investigation and interpretation of the optical properties of CNTs, as well as the identification of CNTs.

(a) Plot the transition energies E_{ii} versus diameter for all nanotubes with diameters ranging from 0.7 to 3 nm and i from 1 to 3.

(b) What is the functional dependence of E_{ii} on the CNT diameter?

5.6. Effective mass comparison between zigzag CNTs and chiral CNTs.

We have frequently stated that the semiconducting zigzag nanotube is a good convenient model for arbitrary semiconducting nanotubes in general. For example, this has been shown to be true in terms of the DOS, which is related to the gradient of the energy dispersion. Let us explore this a bit further for the effective mass parameter, which is related to the curvature of the dispersion.

(a) Numerically compute the effective mass for semiconducting chiral nanotubes for diameters from 1 to 3 nm.

(b) Compare the numerically computed effective mass with the analytical effective mass for semiconducting zigzag CNTs of the same diameter. Quantify any discrepancy in percentage terms.

6 Ideal quantum electrical properties

It seems that the fundamental idea pertaining to quanta is the impossibility to consider an isolated quantity of energy without associating a particular frequency to it.

Louis de Broglie (postulated electron waves)

6.1 Introduction

The goal of this chapter is to explore the excitation and motion of electron waves under ideal conditions in a metallic conductor. By ideal conditions, we mean that electrons can be excited and transported without any scattering or collision involved. The excitation of electrons can be achieved by applying an external potential to energize the electron waves to oscillate more frequently, which can result in a net electron motion in the presence of a driving electric field, say between two ends of a metallic conductor. It is advisable to commit to memory that the absence of electron scattering is technically called ballistic transport; as such, the metallic conductor in this case would be referred to as a ballistic conductor.

Electrically, the ideal excitation and motion of electrons in low dimensions, such as in 1D space, is manifest in the form of a quantum conductance, quantum capacitance, and kinetic inductance, which represents a different paradigm from our classical electrostatic and magnetostatic ideas. The conductance and inductance reflect the electrical properties of traveling electron waves which lead to charge transport and energy storage, while the quantum capacitance accounts for the intrinsic charge storage that comes about from exciting electrons with an electric potential. In macroscopic bulk metals, the quantum electrical properties are not readily observable or accessible owing to the large number of mobile electrons at hand and the frequent collisions involved.[1] Consequently, these quantum electrical properties have only attracted significant interest and scholarship over the last three decades, in part due to manufacturing advancements and innovation in fabricating

---

[1] We will elucidate on how the quantum electrical properties scale from nanoscopic to macroscopic materials as we expand the discussion throughout the chapter.

nanoscale structures and synthesizing nanomaterials. These nanomaterials include nanowires, CNTs, and more recently graphene.

In order to develop our intuition, we will start simply by employing conventional analysis to derive the quantum electrical properties in a 1D conductor to elucidate the salient features and then subsequently apply the derivation to carbon nanomaterials. In Section 6.10 we bring to light a universal relation (vis-à-vis Planck's postulate) by which the quantum conductance of particle-waves can be derived in a more fundamental manner. The universal relation proves to be powerful, as will be seen in Chapter 8 when it is applied to describe charge transport in ballistic CNT transistors. A more general study of carbon interconnects, including length and voltage-dependent scattering, is presented in Chapter 7.

6.2   Quantum conductance

The quantum conductance/resistance represents a different paradigm from our classical ideas regarding resistance, where electrons are seen as mobile particles that experience frequent collisions with the lattice. These collisions are what are responsible for the classical resistance captured by Ohm's law. It follows that, in the absence of scattering, electrons should be able to travel pleasantly through the conductor and hence the resistance would vanish to zero or, alternatively, the conductance becomes infinite. Is this really the case? To address this question and gain insight into the fundamental origin of conductance in a scattering-free metal, let us consider a 1D ballistic wire or channel connected to reservoirs of electrons, as shown in Figure 6.1. In practice, the reservoirs are realized as bulk metals that have cross-sectional dimensions that are orders of magnitude greater than the channel. Ideally, electrons experience no reflection, as they couple from the reservoir into the channel and vice versa; as such, the reservoir is called a transparent contact.

The current flowing through the ballistic channel from electron waves with an energy E can be determined from the flux equation defined as

$$I(E) = \frac{\partial q(E)}{\partial t} = \frac{\partial q(E)}{\partial l}\frac{\partial l}{\partial t}, \tag{6.1}$$

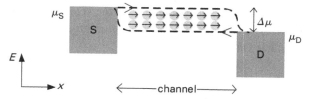

Fig. 6.1   Band diagram of a ballistic channel connected to source (S) and drain (D) reservoirs of electrons. The arrows indicate the direction of electron flow from both reservoirs. The net current is determined by the flux of carriers within the energy range of $\Delta\mu$.

where $\partial q/\partial l$ is the charge density per unit length and $\partial l/\partial t = v(E)$, the velocity of an electron with energy E. The velocity is a vector quantity which results in some of the charge traveling with positive velocity and some with negative velocity. Let us label the charge traveling with positive velocity q^+ and the charge traveling with negative velocity q^-. It follows that the net current from electrons with energies between E and $E + dE$ is

$$I(E)dE = \left(\frac{\partial q^+(E)}{\partial l} - \frac{\partial q^-(E)}{\partial l}\right) v(E)dE. \tag{6.2}$$

Note that the positive term represents right-moving carriers or forward current, while the negative term is due to left-moving carriers or reverse current. At equilibrium, the number of charges traveling with $+v(E)$ equals the number of charges traveling with $-v(E)$, leading to zero net current. The charge density is related to the DOS $g(E)$ via the relations

$$\frac{\partial q^+(E)}{\partial l}dE = e\frac{g(E)}{2}N_{\text{ch}}(E)F(\mu_S) \, dE, \tag{6.3}$$

$$\frac{\partial q^-(E)}{\partial l}dE = e\frac{g(E)}{2}N_{\text{ch}}(E)F(\mu_D) \, dE, \tag{6.4}$$

where the factor of 2 is due to half of the states at E having a positive velocity (right-moving states) and the other half having negative velocity (left-moving states). $N_{\text{ch}}(E)$ is the total number of propagating channels or modes that are degenerate at energy E and typically includes the number of degenerate subbands. F is the Fermi–Dirac distribution given by

$$F(\mu) = \frac{1}{1 + e^{(E-\mu)/k_B T}}, \tag{6.5}$$

where k_B is Boltzmann's constant and T is the temperature. We employ chemical potential μ instead of the Fermi energy in these calculations mainly to emphasize the departure from the equilibrium E_F. The total net current is

$$I = \int_{-\infty}^{\infty} I(E) \, dE = e \int_{-\infty}^{\infty} \frac{g(E)}{2}N_{\text{ch}}(E)v(E)[F(\mu_S) - F(\mu_D)] \, dE, \tag{6.6}$$

where μ_S and μ_D are the chemical potentials at the source and drain reservoirs respectively. The charge velocity of right-moving or left-moving states is related to the density of states via (see Eq. (5.49))

$$v(E) = \frac{1}{\hbar\pi g(E)/2} = \frac{4}{hg(E)}, \tag{6.7}$$

where h ($\hbar$) is Planck's (reduced Planck's) constant. Considering low-energy transport where N_{ch} is a constant within the energy range of interest, the net current simplifies to

$$I = \frac{2e}{h} N_{ch} \int_{-\infty}^{\infty} [F(\mu_S) - F(\mu_D)] \, dE. \tag{6.8}$$

The factor of 2 in the numerator is due to spin degeneracy.[2] The chemical potentials are controlled by the applied drain to source voltage V, which can be written in a symmetrical manner as $\mu_S = E_F + eV/2$, $\mu_D = E_F - eV/2$, and $\Delta\mu = \mu_S - \mu_D = eV$. E_F is the equilibrium Fermi energy, which can be arbitrarily set to zero for convenience. The solution to the integral results in an interestingly simple expression for the ballistic current:

$$I = \frac{2e}{h} N_{ch} \Delta\mu = \frac{2e^2}{h} N_{ch} V. \tag{6.9}$$

The ballistic current reveals that current is determined by the difference in chemical potentials $\Delta\mu$, reflecting the states with a positive velocity that have been excited courtesy of the external potential. The quantum conductance G_q is

$$G_q = \frac{dI}{dV} = \frac{2e^2}{h} N_{ch} = G_o N_{ch}, \tag{6.10}$$

where $G_o = 2e^2/h$ is a basic unit of quantum conductance. Numerically, $G_o \approx 77.5 \, \mu S$ or alternatively in terms of resistance $1/G_o \approx 12.9 \, k\Omega$. This result was initially a surprise to many because it asserts that the resistance of an ideal scattering-free 1D conductor is finite and in fact relatively large (on the order of kilo-ohms). It has since become widely recognized that this two-terminal resistance represents a fundamental *contact* resistance between the reservoirs and the ballistic channel. This contact resistance can be interpreted as a fundamental transmission coefficient describing how electrons couple from the reservoir into the 1D channel and vice versa. As an example, consider an electron traveling from the source at a chemical potential μ_S. When the electron reaches the drain end, it will inevitably have to scatter at the drain reservoir in order for the drain chemical potential to be maintained at μ_D. This is the origin of the quantum resistance. Nonetheless, if we probed the voltage drop in the ballistic channel in a non-invasive way, should the measured voltage drop not be zero? If the measured voltage drop is finite then this would be contrary to the ballistic nature of the channel. Fortunately, de Picciotto et al.,[3] have been able to provide an experimental answer to this question by performing four-terminal measurements of a ballistic channel, confirming that the

---

[2] It might not be obvious how the factor of 2 accounts for spin degeneracy. To see the origin of 2, the reader should recall from Chapter 5 that spin degeneracy was included in the definition of the density of states.

[3] R. de Picciotto, H. L. Stomer, L. N. Pfeiffer, K. W. Baldwin and K. W. West, Four-terminal resistance of a ballistic quantum wire. *Nature*, **411** (2001) 51–4.

voltage drop inside the channel indeed vanishes to zero consistent with the band diagram in Figure 6.1, where the bands are flat within the channel.

The ballistic condition, which reveals a length-independent conductance, applies when the device or nanomaterial dimensions are much less than the mean free path[4] l_{mfp} of electron transport. There are a variety of scattering mechanisms that determine the mean free path, which will be discussed in subsequent chapters. In the meantime, we can reach a broad conclusion that, in a narrow device with a length $l \gg l_{mfp}$, electrons will experience frequent scattering with the lattice, yielding a more general expression for the quantum conductance:

$$G_q = \frac{2e^2 N_{ch}}{h} T, \tag{6.11}$$

where T can be considered the transmission coefficient or probability of electrons propagating through the channel without scattering and satisfies the condition $0 < T < 1$. $T = 1$ is the ballistic condition. Equation (6.11) is widely known as Landauer's formula after Rolf Landauer, who pioneered the development of the expression.[5] Scaling from one to three dimensions, the number of modes increases by several orders of magnitude, resulting in a vanishingly small resistance. In any dimension, inclusion of the effects of electron scattering by the lattice leads to a resistance that depends on the length of the conductor and the recovery of Ohm's law.

Additional insights regarding the quantum conductance

Let us return to the ballistic limit and focus on N_{ch} for the moment to gain further understanding. As mentioned earlier, $N_{ch}(E)$ is the number of degenerate subbands at energy E, sometimes called the number of propagating channels or modes. It is worthwhile noting that N_{ch} is the only variable in the quantum conductance representing the material, temperature, and field dependence. For the lowest subband in metallic CNTs, $N_{ch} = 2$, resulting in $R_q \approx 6.45\,k\Omega$. In principle, the discrete nature of the quantum conductance is observable at very low temperatures ($T \to 0$ K) by populating subbands one at a time with an external voltage. An experimental graph of the discrete step-like nature of the quantum conductance observed in a novel three-terminal device known as a quantum point contact is shown in Figure 6.2. As temperature increases from 0 K, the steps become increasingly blurred because the exponential tail of the Fermi–Dirac distribution includes otherwise higher subbands, leading to a continuum of the quantum conductance as a function of applied voltage.

---

[4] The mean free path is the average distance electrons travel in a conductor before experiencing a collision or scattering event.

[5] Y. Imry and R. Landauer, Conductance viewed as transmission. *Rev. Mod. Phys.*, **21** (1999) 5306–12.

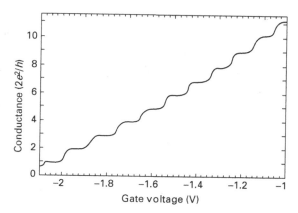

Fig. 6.2 Conductance of a quantum point contact (at $T = 0.6$ K), revealing the quantized nature of conductance in a narrow constriction. Adapted with permission from B. J. van Wees *et al.*, *Phys. Rev. Lett.*, **60**, (1988) 848–50. Copyright (1988) by the American Physical Society.

Additional insights are available from a more detailed study of the quantum conductance. In metallic CNTs at very low energies and finite temperatures (including room temperature) with $l \ll l_{mfp}$, it follows that the exponential tail of the Fermi–Dirac distribution in the second subband leads to negligible contribution of current and conductance from the second subband compared with the first subband. This implies that the quantum conductance is determined only by the first subband with negligible temperature dependence. Hence, measurement of the quantum conductance is a very valuable method for characterizing the quality of the contacts to the nanotubes under different fabrication process conditions. The mean free path at low energies in high-quality CNTs is about 1 μm at room temperature. On the whole, this characterization method is extendable to graphene ribbons and other metallic nanomaterials, provided the device dimensions are much less than the mean free path. For the case of non-ideal or non-transparent contacts, an additional parasitic or residual contact resistance must be added in series to the quantum resistance.

An important insight arises when we include electron–electron interactions in an otherwise ballistic channel that is connected to reservoirs of independent electrons.[6] In an ideal 1D channel where electron–electron interactions are viewed as a Luttinger liquid,[7] Maslov and Stone[8] have shown that the quantum conductance is not at all modified by such electron interactions. Fortunately, we

---

[6] Practical reservoirs achieved with common metals are for the most part reservoirs of independent (non-interacting) electrons, often called a Fermi gas. As a primer, systems of non-interacting electrons are termed gases, and systems of interacting electrons are termed liquids or fluids.

[7] Luttinger liquid interactions are peculiar to 1D channels, and are characterized by charge density and spin density waves propagating at different velocities, otherwise known as spin–charge seperation.

[8] D. L. Maslov and M. Stone, Landauer conductance of Luttinger liquids with leads. *Phys. Rev. B*, **52** (1995) R5539–42.

can understand why this is the case by straightforward reasoning without having to consider the full quantum mechanical mathematical proof. The reason is because, as mentioned earlier, the quantum conductance is fundamentally due to the dynamics of electrons arising from the reservoir and propagating into the channel. Since the electrons from the contact reservoirs are independent electrons, charge transport and conductance is due to the propagation of independent electrons.

So far we have attributed the quantum conductance as arising from reservoir contacts to the 1D channel. However, the role of the contacts is not at all obvious and certainly not explicit in the derivation of the quantum conductance. In that sense, attributing the quantum conductance to coupling between the contacts and the nanotube is somewhat of an ad hoc argument. Perhaps if we take a closer look at the expression for the quantum conductance (Eq. (6.10)), it might be possible to incorporate the contacts–nanotube coupling directly into the derivation and indeed show that it is this coupling that is responsible for the quantum conductance. For the simplest case of $N_{ch} = 1$, Eq. (6.10) reveals to us that it is essentially two electrons that are responsible for current per unit time or period. Therefore, we can model the channel as containing a single state or level at a specific energy that holds two electrons. Datta[9] has shown mathematically that, when the channel is coupled to the reservoirs, the single state inevitably broadens and the quantum conductance falls out naturally in the derivation as a consequence of coupling between the nanotube and the reservoir contacts.

It is tempting to apply the quantum conductance to graphene immediately. However, this is not as straightforward as it might seem. Consider a large sheet of graphene (say greater than $> 100\ \mu m \times 100\ \mu m$); electrons will inevitably scatter as they travel within this 2D sheet because typical electron mean free paths are significantly much less than the sheet dimensions. As a result, the conductance or resistance is of a classical type and can be described using Ohm's law using concepts such as the mean free path, mobility, or conductivity. To see the quantum conductance in graphene, the length and width of the graphene have to be much smaller than the mean free path, which implies that it essentially has be reduced to one dimension in the form of a nanoribbon.

6.3   Quantum conductance of multi-wall CNTs

Multi-wall CNTs are composed of multiple concentric shells of single-wall nanotubes with an outer diameter that is anywhere from several nanometers to about a 100 nm. The inner diameter is the diameter of the smallest shell d_{in}, the outer

---

[9] S. Datta, Quantum Transport: Atom to Transistor (Cambridge University Press, 2005), Chapter 1.

diameter is the diameter of the largest shell d_{out}, and the shell-to-shell spacing is approximately equal to the interlayer spacing in graphite ($\delta \sim 0.34$ nm). In the simplest model, the walls or shells of an MWCNT are considered to be non-interacting, which implies that each shell can be treated as an independent single-wall nanotube. Therefore, the electrical conductance of an MWCNT is a linear sum of the conductance of each shell. In general, multi-wall nanotubes consisting of more than three shells exhibit metallic properties. One can arrive at this conclusion from a variety of arguments of varying sophistication. For example, based on the standard statistical distribution of single-wall nanotubes, one-third of all the shells are metallic and, hence, MWCNTs with greater than three shells have a metallic character overall. Even the semiconducting shells can be quasi-metallic at room temperature,[10] if the diameter of the shell d_s is large enough such that the first subband is sufficiently populated with electrons excited by the thermal energy.

In the context of electron transport, an ideal multi-wall nanotube is a conductor where the reservoir contacts make identical transparent connections to all the shells[11] and electrons flow through the conductor without scattering. These idealistic simplifications are necessary to set the stage in obtaining basic insights regarding ballistic transport in multi-wall nanotubes. Fortunately, the insights are of a general nature that provides understanding of single-shell and multi-shell conductance. Without much ado, the quantum conductance of a multi-wall nanotube $G_{q,mw}$ can be written as the derivative of the total ballistic current (adapted from Eq. (6.17)):

$$G_{q,mw} = \sum_{d_s} \frac{\partial I}{\partial V} = \alpha \frac{2e}{h} \sum_{d_s} \frac{\partial}{\partial V} \left[\sum_{j} \int_{-\infty}^{\infty} (F(\mu_S) - F(\mu_D)) \; dE \right], \quad (6.12)$$

where d_s is to be summed over all the shells or walls in the multi-wall nanotube in steps corresponding to 2δ; j is the subband index, and the inner sum is over the subbands of each shell. We recall from Chapter 5 that α accounts for the Brillouin zone symmetry or degeneracy. Considering achiral nanotubes: $\alpha = 1$ if the wall or shell has a zigzag character, because the bottoms of the subbands are all located at the center of the zone, and $\alpha = 2$ for armchair shells, because there are two locations for the band bottoms, $k \sim \pm 2\pi/3a$, where k is the CNT Brillouin zone

---

[10] Quasi-metallic nanotubes refer to nanotubes that possess a small bandgap that is of the order of $k_B T$. As such, the first subband is sufficiently populated by thermally excited electrons and the CNT behaves like a metal (linear current–voltage relation).

[11] It has so far proved challenging to routinely fabricate transparent contacts to all the shells of an MWCNT. Additionally, it is by no means a trivial matter to even determine the number of shells. Nonetheless, it is worthwhile to study the ideal properties.

wavevector and a is the lattice constant ($a \sim 2.46$ Å). For electrons (holes) in the jth-subband, the energy at the bottom (top) of the subband is $E_j(-E_j)$, which can be incorporated into the limits of the integral.

$$
G_{q,mw} = \alpha G_o \sum_{d_S} \sum_j \frac{\partial}{e \partial V} \left[\int_{-\infty}^{-E_j} (F(\mu_S) - F(\mu_D)) \, dE \right.
$$

$$
\left. + \int_{E_j}^{\infty} (F(\mu_S) - F(\mu_D)) \, dE \right]. \tag{6.13}
$$

The first term in the integral is the contribution from holes and the latter term is the contribution from electrons. The solution to the definite integral leads to

$$
G_{q,mw} = \alpha \sum_{d_S} \sum_j \left[1 + \frac{1}{1 + e^{(eV+2E_j)/2k_B T}} \right.
$$

$$
\left. - \frac{1}{1 + e^{(eV-2E_j)/2k_B T}} \right] G_o = N_{ch} G_o, \tag{6.14}
$$

where N_{ch} is the total number of channels the MWCNT offers (the coefficient of G_o). Equation (6.14) is a general formula for the quantum conductance of multi-wall nanotubes with shells of arbitrary chirality. It can also be applied to determine the conductance of quasi-metallic single-wall nanotubes. It is notable that the conductance is, on the whole, voltage dependent. In the ballistic regime, higher voltages increase the contribution from higher subbands, resulting in a larger conductance. For the first subband of a metallic shell ($E_j = 0$), the conductance reduces to the expected value of αG_o.

To make further progress that explicitly expresses E_j in terms of the subband index j and the shell diameter, it is worthwhile exploring the conductance of a multi-wall nanotube with only armchair shells because of its high-symmetry which simplifies the analysis. The total number of channels $N_{ch,amw}$ of an armchair MWCNT is an upper bound on the ballistic conductance achievable from multi-wall nanotubes of arbitrary chirality:

$$
N_{ch,amw} = 2 \sum_{d_S,j=n} 1 + 2 \sum_{d_S} \sum_{j \neq n} \left[1 + \frac{1}{1 + e^{(eV+2E_j)/2k_B T}} - \frac{1}{1 + e^{(eV-2E_j)/2k_B T}} \right].
$$

$$
\tag{6.15}
$$

The number of channels has been organized into two parts. The first part is the contribution from the lowest subband (the lowest subband has an index $j = n$), while the latter part is the contribution from the higher subbands. In the limit of small voltages ($V \rightarrow 0V$) applicable to practical interconnects, it follows that

$$
\begin{aligned}
N_{\text{ch,amw}} &= 2 \left[N_{\text{sh}} + \sum_{d_s} \sum_{j \neq n} \frac{2}{1 + e^{E_j/k_B T}} \right] \\
&= 2 \left[N_{\text{sh}} + 2 \sum_{d_s} \sum_{j \neq n} F(E_j) \right],
\end{aligned}
\tag{6.16}
$$

where N_{sh} is the number of shells. $F(E_j)$ is the Fermi–Dirac function evaluated at an energy equal to E_j, and can be interpreted as the weighted contribution of the jth-subband.

Moving forward, we desire a general expression for the energies of the bottom of the lowest subbands as a function of shell diameter. Even though the shell is assumed to have an armchair chirality for mathematical simplicity, the final results generally apply to shells of arbitrary chirality essentially because the analysis is exploring local properties around the K-point of the Brillouin zone of graphene, which is largely chirality independent. From Chapter 4, the energy minima of the jth-subband of an armchair nanotube is

$$
E_j \approx \frac{\gamma \pi}{n} |j - n|.
\tag{6.17}
$$

For the first-subband, $j = n$ and we obtain the expected result of $0\,\text{eV}$. The subbands next to (but not at) the K-point of graphene always have a twofold degeneracy with one set of subbands corresponding to values of $j > n$, while the other subbands in the twin pairs have values of $j < n$. With this in mind, it is easiest to focus on the positive branch of Eq. (6.17) and introduce a factor of two for the subband degeneracy afterwards:

$$
\begin{aligned}
E_j &\approx \frac{\gamma \pi}{n}(j - n), \\
E_{j'} &\approx \frac{\gamma \pi}{n} j', \quad j' = 1, 2, \ldots, n,
\end{aligned}
\tag{6.18}
$$

where $(j - n)$ is replaced with the symbol j' and corresponds to the number of additional subbands that provide current. It is important to keep in mind that Eq. (6.18) is very accurate only for the lowest subbands, which in fact is where our interest lies. The increasing errors incurred in Eq. (6.18) for much higher subbands are made negligible by the rapid exponential decline of the Fermi–Dirac function. $E_{j'}$ can be explicitly expressed in terms of the shell diameter via the diameter–chirality relation for armchair nanotubes, $n = \pi d_s (\sqrt{3}a)^{-1}$:

$$
E_{j'}(d_s) = \frac{\sqrt{3}\alpha\gamma}{d_s} j'.
\tag{6.19}
$$

Numerically, $E_{j'} \sim 1.3j'(\text{eVnm})/d_s(\text{nm})$. The diameter-dependent Eq. (6.19) can be substituted for E_j in order to evaluate the number of channels in an armchair MWCNT:

$$N_{\text{ch,amw}} = 2\left[N_{\text{sh}} + 4\sum_{d_s}\sum_{j'=1}^{n}\frac{1}{1 + e^{E_{j'}/k_B T}}\right], \tag{6.20}$$

where the sum over the subbands will largely be determined by roughly the first 10-15 lowest subbands at room temperature and, as such, the sum need not extend to the highest subbands close to $j' = n$. The additional factor of two in the latter term is to account for the subband degeneracy.

Figure 6.3 plots Eq. (6.20) as a function of the diameter for shells with an armchair (metallic) chirality. Also shown in the plot is N_{ch} for semiconducting shells and a standard average value for the number of channels. The standard average value of the number of channels $N_{\text{ch,avg}}$ is calculated by weighting the contributions of a metallic shell and a semiconducting shell of the same diameter by 1/3 and 2/3 respectively and is a more relevant value which we will use from now onwards, since it reflects the random distribution of nanotube chiralities. Judging from Figure 6.3 it is possible to obtain an accurate linear approximation for the average number of channels per shell at room temperature:

$$N_{\text{ch,avg}}(d_s) \approx a_1 d_s + a_2, \tag{6.21}$$

where $a_1 \sim 0.107\text{nm}^{-1}$, and $a_2 \sim 0.088$, with good accuracy for practical diameters of interest (error < 10% for d_s > 6nm, and error <2% for $d_s \geq 9.5$ nm). At smaller diameters, $N_{\text{ch,avg}}$ approaches a constant value of 2/3.

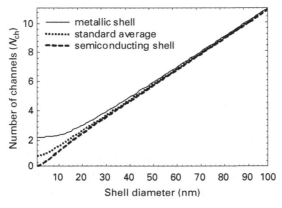

Fig. 6.3   Number of conducting channels of a multi-wall nanotube as a function of the shell diameter at room temperature in the limit of a small applied voltage. The standard average is 1/3 (metallic shell) + 2/3 (semiconducting shell), which reflects the standard statistical distribution of nanotube chiralities, is also shown. $\gamma = 3.1$ eV for this plot.

The total number of channels for a multi-wall nanotube is the sum of the average channels per shell:

$$N_{\text{ch,avg}} = \sum_{d_s} N_{\text{ch,avg}}(d_s) = \sum_{i=0}^{N_{\text{sh}}-1} [a_1(d_{\text{in}} + 2\delta i) + \alpha_2], \qquad (6.22)$$

where $d_{\text{in}} + 2\delta i$ represents the diameter of the ith-shell. Algebraic simplification of the finite series leads to

$$N_{\text{ch,avg}} = \sum_{i=0}^{N_{\text{sh}}-1} (a_1 d_{\text{in}} + a_2) + 2\delta a_1 \sum_{i=0}^{N_{\text{sh}}-1} i = N_{\text{sh}} (a_1 d_{\text{in}} + a_2) + \delta a_1 N_{\text{sh}}(N_{\text{sh}}-1).$$

$$(6.23)$$

Substituting for N_{sh} ($N_{\text{sh}} = 1 + (d_{\text{out}} - d_{\text{in}})/2\delta$) yields a closed-form formula:[12]

$$N_{\text{ch,avg}} = \left(\frac{d_{\text{out}} - d_{\text{in}}}{2\delta} + 1 \right) \left(a_1 \frac{d_{\text{out}} + d_{\text{in}}}{2} + a_2 \right). \qquad (6.24)$$

Eq. (6.24) is the central result of this section, and will be employed in the discussion of the resistance of MWCNT interconnects in Chapter 7.

6.4 Quantum capacitance

The phrase *quantum capacitance* is relatively new, but the concept is an old one.[13] The quantum capacitance is a model of the intrinsic charge storage which is excited by a small-signal electric potential. Fundamentally, it arises from the finite energy needed to put more free carriers in the system because of the quantized energy levels which have a finite number of states. The quantum capacitance is generally not accessible (i.e., $C_q \to \infty$) if (i) the DOS is very large or (ii) the energy level separation between states is vanishingly small. The quantum capacitance is often very large in bulk 3D conductors because of the great number of mobile electrons ($\sim 10^{22}$–10^{23} cm^{-3}) which are excitable by a relatively small amount of energy. In an actual circuit or device, such as a parallel plate capacitor, the quantum capacitance is often in series with the geometric electrostatic capacitance. As a result

---

[12] Strictly speaking, N_{sh} can only take on integer values, and as such the expression for N_{sh} should be rounded to an integer. However, if the number of shells is large, then there is little error in allowing N_{sh} to be continuous for analytic convenience. For example, the error between $N_{\text{sh}} = 10$ and $N_{\text{sh}} \sim 10.5$ is approximately 5%.

[13] The phrase quantum capacitance was coined in: S. Luryi, Quantum capacitance devices. *Appl. Phys. Lett.*, **52** (1988) 501–3. Historically, the concept has been referred to as the DOS capacitance in semiconductor physics.

of the relatively large value of the quantum capacitance in bulk metallic conductors, the electrostatic capacitance dominates. Fortunately, in reduced dimensions, the relatively fewer number of electrons which require larger excitation energies makes the quantum capacitance smaller and, hence, more accessible. However, it is important to note that C_q is not unique to CNT or 1D systems. It is just that 1D systems have a low DOS, and, therefore, are more *visible* electrically. Formally, the quantum capacitance (per unit length in one dimension and per unit area in two dimensions) is

$$C_q(\varphi_s) = \frac{-\partial q_\ell}{\partial \varphi_s} = -\frac{\partial \mu}{\partial \varphi_s}\frac{\partial q_\ell}{\partial \mu} = -e\frac{\partial q_\ell}{\partial \mu}, \tag{6.25}$$

where φ_s is the local electrostatic channel or surface potential given by $e\varphi_s = \mu$, and q_ℓ is the charge density given by the integral of the occupied DOS with energy (charge per unit length in one dimension and per unit area in two dimensions). Let us consider electronic nanomaterials such as CNTs and graphene that possess electron–hole symmetry about 0 eV. In that case, the charge density is

$$q_\ell = e\int_{-\infty}^{0} g(E)[1 - F(E,\mu)]\, dE - e\int_{0}^{\infty} g(E)F(E,\mu)\, dE, \tag{6.26}$$

where the former and latter integrals are the contribution of holes and electrons respectively. Substituting Eq. (6.26) into Eq. (6.25) produces

$$C_q = -e^2\frac{\partial}{\partial \mu}\int_{-\infty}^{0} g(E)[1 - F(E,\mu)]\, dE + e^2\frac{\partial}{\partial \mu}\int_{0}^{\infty} g(E)F(E,\mu)\, dE,$$

$$C_q = e^2\int_{-\infty}^{0} g(E)F_{th}(E,\mu)\, dE + e^2\int_{0}^{\infty} g(E)F_{th}(E,\mu)\, dE,$$

$$C_q = e^2\int_{-\infty}^{\infty} g(E)F_{th}(E,\mu)\, dE, \tag{6.27}$$

where

$$F_{th} = \frac{\partial F}{\partial \mu} = \frac{1}{4k_BT}\text{sech}^2\left(\frac{E - \mu}{2k_BT}\right) \tag{6.28}$$

is the thermal broadening function and has the form of a bell-shaped curve with a full-width at half-maximum that broadens with temperature. The hyperbolic secant is defined as $\text{sech} = 2/(e^x + e^{-x})$. There are two insights we gain courtesy of the thermal broadening function:

(i) Since F_{th} is centered at the Fermi energy or chemical potential, the states about μ contribute most to the quantum capacitance; hence C_q is a strong local function of the DOS around μ. Another way to look at is that F_{th} serves as a selection window or filter for states about μ, and suppressing the contribution of states farther away from μ.

(ii) By electrostatically controlling the position of μ (say by a gate voltage), we are able to move the broadening function around, and for the argument given in (i) we observe that the graphical profile of the quantum capacitance will in general reflect the profile of the DOS.

Eq. (6.27) can be employed as a starting point for capacitance calculations for electronic materials that possess electron–hole symmetry. We note that $T = 0$ K, represents a special condition that requires the use of the Dirac delta function in place of the thermal broadening function in Eq. (6.27). At this temperature, which is commonly referred to as the ground state of the solid, the quantum capacitance simplifies to a value determined by the DOS at the Fermi energy:

$$C_q(0, K) = e^2 \int_{-\infty}^{\infty} g(E)\delta(E - E_F) \, dE = e^2 \, g(E_F). \qquad (6.29)$$

Up until now it has not been clear whether we can measure or make use of the quantum capacitance in an electronic circuit. Indeed, the capacitance we will observe in a circuit is always a combination of the quantum capacitance with an external capacitance. This is because capacitance is defined with respect to two terminals or electrodes and, hence, we must consider electric fields between the two electrodes, which inevitably results in an electrostatic capacitance C_{es}. To see this more clearly, consider the simple model of a nanowire (a wire with a width of the order of nanometers) over a ground plane as shown in Figure 6.4.

By applying a small signal potential v_{ac} riding on top of a static voltage V_{dc} we can compute the total capacitance C_{tot} between the two electrodes. Notice that

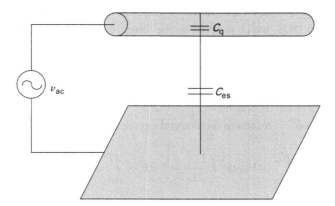

Fig. 6.4   A simple capacitive model of a nanowire over a ground plane excited by a voltage source v_{ac}. The total capacitance in the circuit is a series combination of the quantum C_q and electrostatic C_{es} capacitances.

since capacitance is defined with respect to a change in potential, C_{tot} is the same regardless of whether we chose the capacitance to be that of a wire over a plane or a plane over a wire. The question that arises is: Will all the applied voltage be across the dielectric between the electrodes? The short answer is no. Some of the voltage will go towards raising the chemical potential and increasing the charge in the nanowire, and the remaining voltage will be across the dielectric representing the electric field between the two electrodes. In analogy to electrical circuits, this is the same situation occurring when two capacitors are connected in series. Therefore, we can model the observed capacitance as shown in Figure 6.4 as a series combination of the quantum capacitance, which accounts for the intrinsic charge storage, and an electrostatic capacitance, which accounts for the electric field:

$$C_{tot}(V_{dc}) = \frac{C_{es}C_q(\varphi_s)}{C_{es} + C_q(\varphi_s)}, \tag{6.30}$$

with the surface potential calculated according to the capacitive voltage divider formula:

$$\varphi_s(V_{dc}) = \frac{C_{es}}{C_{es} + C_q(\varphi_s)} V_{dc}. \tag{6.31}$$

The electrostatic capacitance is computed as usual from the electric fields between the wire and the plane. In general, the coupled equations of (6.30) and (6.31) have to be solved self-consistently to obtain the correct solution.

6.5   Quantum capacitance of graphene

The low-energy quantum capacitance for graphene can be calculated directly from its low-energy DOS, which is given by the linear relation (from Eq. (3.42))

$$g(E) = \frac{2}{\pi(\hbar v_F)^2}|E| = \beta_g|E|, \tag{6.32}$$

where β_g is a material constant ($\beta_g \sim 1.5 \times 10^6 \, \mu m^{-2} eV^{-2}$). Substituting the DOS into Eq. (6.27) results in the integral expression

$$C_q = \frac{\beta_g e^2}{4k_B T}\left[\int_{-\infty}^{0} -E \mathrm{sech}^2\left(\frac{E-\mu}{2k_B T}\right) dE + \int_{0}^{\infty} E\,\mathrm{sech}^2\left(\frac{E-\mu}{2k_B T}\right) dE\right]. \tag{6.33}$$

For graphene conductors or interconnect applications, the intrinsic or equilibrium quantum capacitance C_{qi} is of primary concern. C_{qi} is determined when the chemical potential or Fermi energy is at equilibrium ($\mu = 0$).

$$C_{qi} = 2\frac{\beta_g e^2}{4k_B T} \int_0^\infty E \operatorname{sech}^2\left(\frac{E}{2k_B T}\right) dE, \tag{6.34}$$

$$C_{qi} = \beta_g e^2 k_B T \ln(4), \tag{6.35}$$

which at room temperature (300 K) is $C_{qi} \sim 0.8\,\mu\text{F cm}^{-2} = 8\,\text{fF}\,\mu\text{m}^{-2}$. Of additional interest is the value and dependence of the quantum capacitance on the chemical potential, at least for small departures from equilibrium. At finite temperatures ($T \neq 0$), the exact solution to Eq. (6.33) is given in terms of the logarithmic function:

$$C_q = \beta_g e^2 \left[2k_B T \ln(1 + e^{\mu/k_B T}) - \mu\right], \tag{6.36}$$

which essentially has a V-shaped profile, as shown in Figure 6.5. For moderate non-zero values of the chemical potential, a simple linear relation between C_q and μ is obtained by examining the instantaneous slope m of Eq. (6.36):

$$m = \frac{\partial C_q}{\partial \mu} = \beta_g e^2 \left[1 - \frac{1}{1 + e^{\mu/k_B T}} + k_B T \ln(1 + e^{\mu/k_B T}) - \mu\right], \tag{6.37}$$

which reduces to for $m \approx \beta_g e^2$ for $\mu > 3k_B T$. It follows, then, that a simple linear relation for the quantum capacitance is

$$C_q = \beta_g e^2 |\mu| \approx 1.5 \times 10^6 |\mu| (\text{F}\,\mu\text{m}^{-2}) \quad \mu > 3k_B T \text{ (eV)}. \tag{6.38}$$

Alternatively, a more esthetically satisfying analytical formula for the quantum capacitance that preserves the electron–hole symmetry and is applicable with

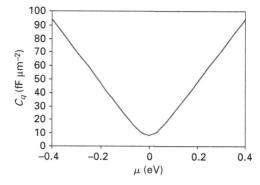

Fig. 6.5 Quantum capacitance of graphene reflecting the V-shaped Dirac profile of the density of states.

reasonable accuracy for arbitrary values of the chemical potential is given by

$$C_q \approx \sqrt{C_{qi}^2 + (\beta_g e^2 \mu)^2}. \tag{6.39}$$

In summary, the key insights regarding the quantum capacitance of graphene include:

(i) The intrinsic quantum capacitance has a minimum value, which at room temperature is, $C_{qi} \sim 8$ fF μm^{-2}.

(ii) The intrinsic C_q is linear with temperature with a slope given by $\beta_g e^2 k_B \ln(4) \sim 0.03$ fF μm^{-2} K^{-1}.

(iii) At moderate energies, the quantum capacitance has a linear dependence on the chemical potential, essentially reflecting the linear profile of the DOS with a slope $\beta_g e^2 \sim 240$ fF μm^{-2} eV^{-1}. Departures from the linear dependence will no doubt occur at high chemical potentials when the dispersion of graphene begins to depart from the Dirac cone. Additionally, departures from the perfectly symmetric V-shaped profile have been observed experimentally due to imperfections in the graphene sheet, where disorder or defects are frequently present.[14]

6.6   Quantum capacitance of metallic CNTs

The quantum capacitance for metallic CNT including only the first subband can be determined from Eq. (6.27) and its DOS, which we recall from Chapter 5 is a material constant:

$$g_o = \frac{8}{\sqrt{3}a\pi\gamma}. \tag{6.40}$$

It follows that the low-energy quantum capacitance is

$$C_q = g_o e^2 \int_{-\infty}^{\infty} \frac{1}{4k_B T} \operatorname{sech}^2 \left(\frac{E - \mu}{2k_B T} \right) \, dE = g_o e^2,$$

$$C_q = \frac{8e^2}{\sqrt{3}a\pi\gamma} = \frac{8e^2}{h v_F} = \frac{4G_o}{v_F}, \tag{6.41}$$

where the relation from Chapter 5, $v_F = \sqrt{3}a\pi\gamma/h$, has been employed to arrive at the latter expression. In contrast to graphene, the metallic CNT quantum capacitance is a material constant reflecting its invariant DOS at low energies. The

[14] J. Xia, F. Chen, J. Li and N. Tao, Measurement of the quantum capacitance of graphene, *Nat. Nanotechnol.*, **4** (2009) 505–9.

factor of 8 accounts for all the degeneracies present in the DOS, including spin and subband degeneracy. Using $v_F \sim 10^8$ cm/s, the quantum capacitance turns out to be $C_q \sim 310 \, \text{aF} \, \mu\text{m}^{-1}$. In interconnect applications, such as a CNT conductor over a ground plane, the quantum capacitance is in general comparable to values of the electrostatic capacitance of a cylindrical wire over a ground plane, implying that the effects of the quantum capacitance are electrically visible and must be considered in metallic nanotube interconnects or transmission lines.[15]

6.7   Quantum capacitance of semiconducting CNTs

Semiconducting CNT offer a more fascinating quantum capacitance profile than metallic nanotubes as a result of the strong energy dependence of their DOS, including the presence of VHSs at the band edge. We recall the simplified DOS of semiconducting nanotubes from Chapter 5 (Eq. 5.20)):

$$g(E,j) \approx \frac{g_0}{2} \frac{|E|}{\sqrt{E^2 - E_{\text{vh}_1}^2}}, \tag{6.42}$$

where j is the subband index and $E_{\text{vh}1}$ is the energy of the VHS. Employing Eq. (6.27), the integral expression to derive the quantum capacitance is

$$C_q = \frac{g_0 e^2}{8k_B T} \left[\int_{-\infty}^{0} \frac{-E}{\sqrt{E^2 - E_{\text{vh}_1}}} \text{sech}^2 \left(\frac{E - \mu}{2k_B T} \right) dE \right.$$
$$\left. + \int_{0}^{\infty} \frac{E}{\sqrt{E^2 - E_{\text{vh}_1}}} \text{sech}^2 \left(\frac{E - \mu}{2k_B T} \right) dE \right]. \tag{6.43}$$

Unfortunately, the integral is not solvable exactly. Certainly, we can proceed by developing some sort of approximation for the integrand that will produce an analytic expression for the quantum capacitance. A better time- and energy-efficient alternative is to return to the starting formula for the quantum capacitance (Eq. (6.25)) and leverage our previous efforts in developing an analytical expression for the charge carrier density. From Chapter 5, the two-subband electron carrier density n_e, which is accurate for values of the chemical potential up to the bottom of the second subband, is

$$n_e \approx \sum_{j=1}^{2} n_e(j) = \sum_{j=1}^{2} \frac{2N_0 \, e^{x_n}}{1 + A e^{\alpha x_n + \beta x_n^2}}, \quad x_n = \frac{\mu - C_{bj}}{k_B T}, \tag{6.44}$$

[15] P. J. Burke, An RF circuit model for carbon nanotubes. *IEEE Trans. Nanotechnol.*, **2** (2003) 55–8.

where C_{bj} is the energy at the bottom of the jth-subband and the semi-empirical fitting parameters A, α, and β are estimated to be $A = 0.63, \alpha = 0.88$, and $\beta = 2.41 \times 10^{-3}$ with good accuracy from $T = 220$ K to $T = 375$ K as discussed in Chapter 5. N_o is the effective DOS and is given by $2N_o = n_i \exp(E_g/2k_B T)$, where n_i is the intrinsic carrier density. Taking the derivative of the electron carrier density according to Eq. (6.25),[16] the quantum capacitance due to electrons is

$$C_{qn} = \sum_{j=1}^{2} \frac{e^2 n_e(j)}{k_B T} \left[1 - \frac{(2N_o\, e^{x_n} - n_e(j))(\alpha + 2\beta x_n)}{2N_o\, e^{x_n}} \right]. \tag{6.45}$$

Similarly, the hole carrier density and evaluated quantum capacitance are respectively

$$n_h = \sum_{j=1}^{2} n_h(j) = \sum_{j=1}^{2} \frac{2N_o\, e^{x_p}}{1 + A\, e^{\alpha x_p + \beta x_p^2}}, \qquad x_p = \frac{V_{tj} - \mu}{k_B T}, \tag{6.46}$$

$$C_{qp} = \sum_{j=1}^{2} \frac{e^2\, n_h(j)}{k_B T} \left[1 - \frac{(2N_o\, e^{x_p} - n_h(j))(\alpha + 2\beta x_p)}{2N_o\, e^{x_p}} \right], \tag{6.47}$$

where V_{tj} is the energy at the top of the jth-subband. The total quantum capacitance for semiconducting nanotubes is a sum of the contributions from electrons and holes:

$$C_q = C_{qn} + C_{qp}. \tag{6.48}$$

In the non-degenerate limit, the quantum capacitances simplify to[17]

$$C_q \approx \frac{4e^2\, N_o}{k_B T} e^{-E_g/2k_B T} \cosh\left[\frac{\mu}{k_B T} \right]; \qquad V_{t1} + 3k_B T < \mu < C_{b_1} - 3k_B T. \tag{6.49}$$

Figure 6.6 plots the quantum capacitance including the electron and hole contributions at room temperature, showing good agreement compared with the more accurate numerical tight-binding computation. It is worth mentioning that the sharp increase in C_q corresponds to the occupation of a subband, and overall the quantum capacitance reflects the peaks and profile of the CNT DOS.

---

[16] Note that chain calculus is needed for the $\delta q_\ell / \partial \mu$ term in Eq. (6.25), i.e.
$\delta q_\ell / \partial \mu = (\delta q_\ell / \partial x_n)(\partial x_n / \partial \mu)$.

[17] D. Akinwande, Y. Nishi and H.-S. P. Wong, Analytical model of carbon nanotube electrostatics: density of states, effective mass, carrier density, and quantum capacitance. *IEEE IEDM Tech. Digest*, (2007) 753.

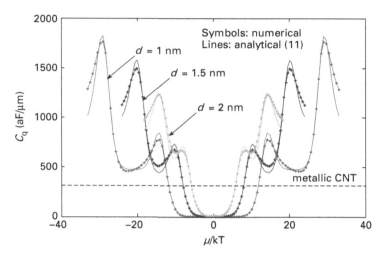

Fig. 6.6   The quantum capacitance of semiconducting CNTs with diameters of 1 nm, 1.5 nm, and 2 nm. Symbols represent numerical tight-binding computation and bold lines are from Eq. (6.48) showing good agreement. The dashed line (included for reference) corresponds to the intrinsic quantum capacitance of metallic CNTs.

6.8   Experimental validation of the quantum capacitance for CNTs

Quite satisfyingly, the fascinating quantum capacitance of semiconducting nanotubes has been confirmed experimentally by Ilani et al.[18] The main results of the experimental measurement corroborate the expected features, including:

(i) voltage-dependent capacitance profile reflecting the DOS;
(ii) the presence of electron–hole symmetry, which has been assumed all along.

A 3D cross-section of the experimental nanotube device is shown in Figure 6.7a. In brief, the experimental device is a two-terminal device with the charge in the nanotube controlled from the electric field from a top-gate voltage V_g. The resulting total capacitance C_{tot} comprises of the series sum of the oxide C_{ox} and quantum capacitances as shown in Figure 6.7b.

$$C_{tot}(V_g) = \frac{C_{ox} C_q(\varphi_s)}{C_{ox} + C_q(\varphi_s)}. \tag{6.50}$$

The oxide capacitance was calculated analytically using formulas for the capacitance of a wire over a plane. The resulting comparison between the experimental data and quantum capacitance model is shown in Figure 6.7c and were performed

---

[18] S. Ilani, L. A. K. Donev, M. Kindermann and P. L. McEuen, Measurement of the quantum capacitance of interacting electrons in carbon nanotubes. *Nat. Phys.*, **2** (2006) 689–91.

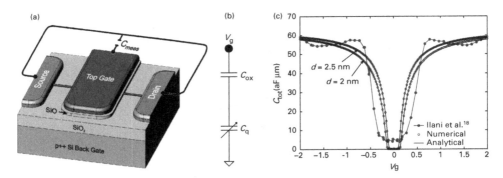

Fig. 6.7 Comparison between quantum capacitance measurement and model predictions performed at 77 K and 1 kHz. (a) A 3D illustration of the experimental top-gated CNT device and (b) equivalent lumped circuit model valid at the measurement temperature and frequency. (c) The total gate capacitance as a function of the top-gate voltage. The analytical model employed Eq. (6.50). An excellent match can be observed between analytical and numerical computation, with both being in strong agreement with the measurement data. (The 3D illustration and experimental data are courtesy of Ilani et al.,[18] and model comparisons courtesy of Liang et al.[19]) Reprinted by permission from Macmillan Publishers Ltd: *Nature Physics*, copyright (2006).

by Liang et al.[19] In general, the analytical capacitance model shows good agreement with the experimental data with the exception of a few discrepancies, including the peak at $V_g \sim \pm 0.8$V observed in the experimental data, which are attributed to interactions among the electrons in the CNT.[18] In addition, at $V_g \sim 0$ V, the measurement is limited by the resolution of the experimental equipment, therefore preventing accurate determination of the capacitance at near-zero voltages.

In addition to the basic physics interest in the voltage-dependent capacitance of semiconducting CNT in elucidating the behavior of electrons in one dimension, the quantum capacitance is also of basic interest in CNT transistors and varactors.[20] In addition, exploiting the quantum capacitance for novel device properties or performance is potentially possible in future nanoelectronics.

6.9 Kinetic inductance of metallic CNTs

The kinetic inductance L_k can be determined from energy considerations. First, let us discuss what is meant by the kinetic inductance. To best understand the kinetic

[19] J. Liang, D. Akinwande and H.-S. P. Wong, Carrier density and quantum capacitance for semiconducting carbon nanotubes. *J. Appl. Phys.*, **104** (2008) 064515.

[20] For example, see D. Akinwande, Y. Nishi and H.-S. P. Wong, Carbon nanotube quantum capacitance for nonlinear terahertz circuits. *IEEE Trans. Nanotechnol.*, **8**, (2009) 51–6; and J. E. Baumgardner, A. A. Pesetski, J. M. Murduck, J. X. Przbysz, J. D. Adam and H. Zhang, Inherent linearity in carbon nanotube field-effect transistors. *Appl. Phys. Lett.*, **91** (2007) 0.52107.

inductance, it is enlightening to recall what the classical magnetic inductance L_m represents. From basic electromagnetics we understand that the motion of charges or current produces a magnetic field which has the ability to do work. The magnetic energy E_m of the magnetic field serves to define a magnetic inductance according to[21]

$$E_m = \frac{1}{2} \int_{\text{allspace}} \mu_m H^2 \, dV = \frac{1}{2} L_m I^2, \qquad (6.51)$$

where H is the magnetic field intensity, μ_m is the material permeability, and dV is the differential volume element. The magnetic inductance reflects the geometry of the conductor and can be seen as a conductor property that models the magnetic energy, a particularly useful property in circuits. However, the magnetic energy is not the only energy that arises from the motion of charges or current. Another manifest energy is the kinetic energy E_k of the electrons, which is essentially where the kinetic inductance originates according to the definition

$$E_k = \frac{1}{2} L_k I^2, \qquad (6.52)$$

where the kinetic energy can be computed either from Newtonian classical mechanics or quantum mechanics. Indeed, the kinetic inductance can be seen in the earliest modern theory of conduction in metals by Drude,[22] where the complex impedance has an imaginary term proportional to frequency, reflecting an inductance from the kinetic energy. The total energy arising from the motion of charges is the sum of the magnetic and kinetic energies, which means that in circuits the kinetic inductance adds in series to the magnetic inductance.

In normal conductors, electrons scatter repeatedly, resulting in dissipation of the kinetic energy. Invariably, whatever kinetic inductance that exists is vanishingly small. In normal metals, therefore, we can conclude that charge transport is of an ohmic nature with the resistance typically dominating the reactance up to optical frequencies. This explains why the kinetic inductance is often not considered in normal electronics. In superconductors, electrons can travel with no loss of their kinetic energy, leading to an appreciable kinetic inductance which is employed in a variety of superconducting devices.[23] In this respect, CNTs are interesting because, while they are not superconductors, they do, however, possess an appreciable kinetic inductance compared with normal bulk metals due to the longer electron

---

[21] For the purpose of refreshing our understanding of L_m, we have recalled the magnetic energy definition of a homogeneous conductor whose material properties are independent of the field or current. The idea of L_m applies to complex conductors as well.

[22] An excellent discussion of Drude's theory of conduction can be seen in, N. W. Ashcroft and N. D. Mermin *Solid State Physics* (Brooks/Cole, 1976) Chapter 1. The original theory was published (in German) by P. Drude in 1900.

[23] For example, a search of *kinetic inductance device* on www.google.com reveals thousands of hits where L_k is utilized in devices ranging from thermometers to resonators to sensors for dark matter.

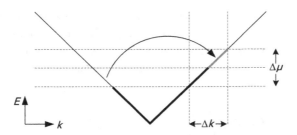

Fig. 6.8   The coupling of energy (via an applied voltage) into a nanotube is akin to the promotion of left-moving carriers into right-moving carriers (gray segment indicated by arrow). The energy stored $\Delta\mu$ by the net carriers (occupying states within Δk) directly manifests as the kinetic inductance. Adapted from M. Bockrath.[24]

mean free path. As a result, it is expected that the kinetic inductance in CNTs will be accessible at more moderate frequencies of the order of $\sim100\,\mathrm{GHz}$, which is potentially useful for high-frequency electronics.

The kinetic inductance of a ballistic metallic CNT can be derived from the net kinetic energy of the carriers that are responsible for the current flow (see Figure 6.8 for a simplified illustration). Assuming the states at the source and drain reservoirs are populated up to μ_S and μ_D respectively, the net total kinetic energy is

$$E_k = \int_0^{\Delta\mu} \frac{g_0}{2} E\,dE = \frac{1}{4} g_0 \Delta\mu^2, \tag{6.53}$$

where $g_0/2$ represents the carriers that provide current and $\Delta\mu$, as mentioned earlier, is the difference in chemical potentials of the source and drain reservoirs, reflecting the energy coupled into the system by the external potential.

The kinetic inductance is determined by equating the kinetic energy to the energy stored in an inductor:

$$E_k = \frac{1}{4} g_0 \Delta\mu^2 \equiv \frac{1}{2} L_k I^2. \tag{6.54}$$

Substituting the current given by Eq. (6.9) and replacing the DOS with the velocity from Eq. (6.7) leads to

$$\frac{1}{4} \frac{4}{h v_F} \Delta\mu^2 = \frac{1}{2} L_k \left(\frac{4e}{h} \Delta\mu \right)^2. \tag{6.55}$$

It follows that the kinetic inductance is

$$L_k = \frac{h}{8e^2 v_F} = \frac{1}{4 G_0 v_F}. \tag{6.56}$$

[24] M. W. Bockrath, Carbon nanotubes: electrons in one dimension. Ph.D. dissertation, University of California, Berkeley, CA, 2002.

With $v_F \sim 10^8$ cm s^{-1}, the kinetic inductance is $L_k \sim 3.2$ nH μm^{-1} (including spin and first subband degeneracy), corroborating initial calculations by Bockrath.[24] This value of the kinetic inductance of metallic CNTs is notably very significant when compared with the magnetic inductance of a metallic wire over a ground plane, which is typically orders of magnitude lower.[15] This implies that the kinetic inductance is potentially useful for high-frequency nanotube circuits. The quantitative (voltage-dependent) value of the kinetic inductance of semiconducting CNTs remains an open question of intellectual and practical interest. In CNT transistor modeling and analysis, L_k is often taken to be of the same order as the kinetic inductance of metallic nanotubes in order to obtain a rough estimate of the impact of the kinetic inductance on charge transport in the nanotube transistor.

In Chapter 7, which focus on metallic interconnects, we will explore the impact of the kinetic inductance in the transmission line model of CNTs. Additionally, the symmetry that exists between the kinetic inductance and quantum capacitance, such as their ratio and their products, which relate to the characteristic impedance and group velocity in the transmission line model, will also be explored in Chapter 7.

6.10 From Planck to quantum conductance: an energy-based derivation of conductance

The derivation of the quantum conductance in Section 6.2 is based on the DOS, but, quite remarkably, the quantum conductance does not depend on the DOS at all. This lack of dependence on the DOS suggests that there might exist a more fundamental method to derive the quantum conductance from basic principles which does not consider the DOS a priori. One line of approach is to start from the most basic postulate in quantum mechanics proposed by Max Planck:[25]

$$E = hf, \tag{6.57}$$

where f is the particle-wave frequency. Since we are more concerned with the flow of electrons per unit period, it is more attractive to rewrite the postulate in terms of electron flux:

$$E = hf = \frac{h}{2e}\frac{2e}{\tau} = \frac{h}{2e}I_E, \tag{6.58}$$

where τ is the electron period and I_E is the current of electrons (including spin) that have an energy E. This current does not take into account the energy degeneracy

---

[25] Max Planck originally proposed the postulate to explain blackbody radiation. Albert Einstein later extended it to explain the photoelectric effect, and Louis de Broglie subsequently generalized the postulate as the basis of his theory of matter waves. All the three ideas based on this single postulate were priceless and, as such, all three pioneers were awarded Nobel Prizes seperately.

that exists at energy E; as such, it is the current per mode:

$$I_E = \frac{2e}{h}E. \tag{6.59}$$

The net current in a 1D channel connected to reservoir contacts is determined by the difference in the chemical potentials of the reservoirs. Assuming all the states are populated up to μ_S at the source reservoir and μ_D at the drain reservoir, with $\mu_S - \mu_D = eV$. The net current I in the channel becomes

$$I = \frac{2e}{h}(\mu_S - \mu_D) = \frac{2e^2}{h}V \tag{6.60}$$

and the quantum conductance per mode falls out naturally as

$$G_o = \frac{2e^2}{h}. \tag{6.61}$$

Similarly, the quantum capacitance of metallic CNTs can also be derived in a straightforward manner from quantum conductance. Let us proceed as follows: the time-varying current in a capacitor is given by the usual displacement current relation:

$$I(t) = \frac{dQ(t)}{dt} = v_F\frac{dQ(t)}{dl}, \tag{6.62}$$

where Q is the total charge in the capacitor and dQ/dl is simply the charge density equal to the product of the capacitance per unit length and the time-varying potential difference $V(t)$:

$$I(t) = v_F C_q V(t). \tag{6.63}$$

Therefore, the quantum capacitance per unit length per mode is

$$C_q = \frac{1}{v_F}\frac{I(t)}{V(t)} = \frac{G_o}{v_F} = \frac{2e^2}{hv_F}, \tag{6.64}$$

which has the same form as Eq. (6.41). It is noteworthy that the quantum capacitance so derived for the metallic nanotube can be extended to the semiconducting nanotube by replacing v_F with the thermally averaged particle velocity. This is a worthwhile exercise for the interested reader seeking to gain more depth.

Likewise, the kinetic inductance of metallic CNT can be derived in a straightforward manner using the quantum conductance:

$$V(t) = \frac{d\phi(t)}{dt} = v_F\frac{d\phi(t)}{dl}, \tag{6.65}$$

where $d\phi/dl$ is the equivalent flux density due to the kinetic inductance and can be expressed as the product of the current and the kinetic inductance per unit length:

$$V(t) = v_F L_k I(t). \qquad (6.66)$$

Therefore, the kinetic inductance per unit length per mode is

$$L_k = \frac{1}{v_F} \frac{V(t)}{I(t)} = \frac{1}{v_F G_o} = \frac{h}{2e^2 v_F}. \qquad (6.67)$$

The successful derivation of the quantum capacitance and kinetic inductance of CNTs from circuit $I-V$ relations and the quantum conductance suggests that G_q not only applies to static current voltage relations, but also applies to alternating current voltage relations.

Noteworthy is that Planck's postulate does not exclusively apply to electrical conductance. Indeed, it can be shown to be useful in deriving both the thermal and spin quantum conductances as well.[26] Moreover, we will employ Planck's postulate (6.58) to derive the ballistic current voltage relationship of a CNT field-effect transistor in Chapter 8.

6.11   Summary

This chapter has explored the electrical properties of quantum mechanical effects in CNTs and graphene. These effects manifest as quantum conductance, quantum capacitance, and the kinetic inductance. We have derived analytical expressions for these electrical properties in the quasi-static limit. Moreover, these electrical models for the quantum mechanical effects utilize the independent-electron or one-electron approximations for understanding the properties of electrons in crystalline solid matter. At low temperatures and under other certain conditions the consequences of electron–electron interactions can become substantial, naturally leading to significant departures from the current electrical models. To first order, these effects can be captured in terms of a renormalized velocity.[27] This

---

[26] The experimental measurement of the quantum thermal conductance was reported by K. Schwab, E. A. Heriksen, J. M. Worlock and M. L. Roukes, Measurement of the quantum of thermal conductance. *Nature*, **404** (2000) 974–7. For theoretical prediction of the spin quantum conductance, see F. Meier and D. Loss, Magnetization transport and quantized spin conductance. *Phys. Rev. Lett.*, **90** (2003) 16724.

[27] For example, see I. Safi, Conductance of a quantum wire: Landauer's approach versus the Kubo formula. *Phys. Rev. B*, **55**, (1997) R7331–4; G. Cuniberti, M. Sassetti and B. Kramer, ac conductance of a quantum wire with electron-electron interactions. *Phys. Rev. B*, **57** (1998) 1515–26; and V. A. Sablikov and B. S. Shchamkhalova, Electron transport in a quantum wire with realistic Coulomb interaction. *Phys. Rev. B*, **58** (1998) 13847–55.

renormalized velocity can be used in place of the Fermi velocity, though experimental corroborations of the renormalization idea are matters of contemporary
research.

Similarly, our treatment of the MWCNTs implicitly assumes that the shells are
non-interacting; that is, electrons in each shell can be treated as independent electrons oblivious of the presence of electrons in the surrounding shells. While this
assumption is expected to hold in the majority of cases, it is always possible to
devise conditions that will violate this assumption and might lead to interesting
physics.

6.12   Problem set

*All the problems are intended to exercise and refine analytical techniques while
providing important insights.* If a particular problem is not clear, the reader is
encouraged to re-study the appropriate sections. Invariably, some problems will
involve making reasonable approximations or assumptions beyond what is already
stated in the specific question in order to obtain a final answer. The reader should
not consider this frustrating, because this is obviously how problems are solved in
the real world.

6.1. Number of channels for metallic single-wall nanotubes.
   The current and conductance of a ballistic single-wall nanotube was derived
   in the text assuming low-energy transport where only the lowest subbands
   contribute appreciable to current. Let us go one step further and actually derive
   the quantum conductance including the contribution of higher subbands while
   still retaining the low-energy approximation.

   (a) Derive the number of channels and quantum conductance for an armchair
       single-wall nanotube including the contributions of higher subbands.
   (b) For an armchair nanotube with diameters of 1 and 2 nm, how much
       larger is the quantum conductance from part (a) normalized to $2G_0$? Is
       the contribution of the higher subbands appreciable or negligible at room
       temperature?
   (c) What is the temperature dependence of the number of channels?
   (d) Likewise, derive the quantum capacitance of metallic single-wall nan-
       otubes including the contribution of higher subbands.

6.2. Voltage dependence of MWCNT conductance.
   The number of channels for MWCNTs was derived in the text in the limit of
   small voltages ($V \rightarrow 0$ V). This was convenient and resulted in an algebraic
   expression for the number of channels for a given multi-wall nanotube that
   can be used to estimate the CNT conductance. However, also of interest is the

voltage dependence of the conductance. Note that the voltage dependence is what is important here, not the field dependence, since the nanotube is in the ballistic mode. This exercise will explore this theme in order to obtain some basic insights.

(a) For a model multi-wall nanotube containing all armchair shells and an inner and outer diameter of 5 nm and 50 nm respectively, plot the number of channels as a function of voltage from 0 V to 5 V at 77 K and at 300 K.

(b) Repeat part (a) for a multi-wall nanotube with an inner and outer diameter of 5 nm and 100 nm respectively.

(c) Based on this simplified model, what conclusions can be made regarding the voltage dependence of the conductance of ballistic multi-wall nanotubes?

6.3. Temperature dependence of MWCNT conductance.
In the limit of small voltages, determine the temperature dependence of the quantum conductance of an ideal multi-wall nanotube from 4 K to 500 K. Plot the temperature dependence for a nanotube with an inner and outer diameter of 10 nm and 50 nm respectively, for the two following conditions:

(a) All the shells of the multi-wall nanotubes make transparent contacts to the electrodes.

(b) Only the outermost shells make transparent contacts to the electrodes. The remaining shells make no electrical contact at all with the electrodes.

(c) Comment on the resulting quantum conductance between the two conditions.

6.4. Non-degenerate quantum capacitance of CNTs.

(a) Show that the non-degenerate quantum capacitance is given by Eq. (6.49).

(b) Plot the error of the non-degenerate quantum capacitance compared with the two-subband analytical quantum capacitance for $E_F = 0$ to $E_F = E_g/2$.

(c) Determine the scaling of the intrinsic quantum capacitance with nanotube diameter, and explain this scaling trend.

6.5. Quantum capacitance of graphene.

(a) Compare the quantum capacitance of graphene (at room temperature) with its electrostatic parallel-plate capacitance for insulator thickness ranging from 5 nm to 1 μm, and a dielectric constant of 4. This comparison can be obtained by superimposing the min/max electrostatic capacitances on the C_q versus μ plot. Comment on this comparison.

(b) Also compare the quantum capacitance of graphene with that of an array of identical metallic single-wall nanotubes. The array can be modeled to consist of 1 nm diameter CNTs with a density of 200 nanotubes/μm. We want to determine which carbon nanomaterial offers the highest intrinsic quantum capacitance. In other words, does the quantum capacitance of an array of parallel single-wall nanotubes approach that of graphene, in the limit of a sheet of CNTs?

7　Carbon nanotube interconnects

If you build a 'better' mousetrap, you'd better know what the existing mousetrap can do.

7.1　Introduction

This chapter explores electron transport in metallic nanotubes as it relates to interconnect applications. Both single-wall and multi-wall CNTs are considered. The reader will find it beneficial to be familiar with Chapter 4, which discusses the structure of nanotubes, and Chapter 6, which explores ideal nanotube electrical properties, such as the quantum conductance, quantum capacitance, and the kinetic inductance. Employing CNTs as metallic wires to route direct-current (DC) and high-speed signals in an integrated circuit was one of the earliest application ideas promoting nanotubes because of their high current-carrying capability and ballistic transport over relatively long lengths. In this light, we will examine both the low-frequency (lossy) and high-frequency (lossless) transmission line models for single-wall and multi-wall nanotubes in order to elucidate their interconnect properties. These models include bias or field-dependent electron scattering in the nanotube vis-á-vis the mean free path. As such, electron scattering and mean free path will be discussed, although at a somewhat elementary level.[1] Additionally, the temperature and diameter dependence of the electron mean free path and resistance will be highlighted.

At the end of the day, future nanomaterials such as CNTs have to be benchmarked against existing materials to quantify any performance benefits over conventional approaches. For this purpose, we will discuss the performance of CNTs compared with copper, which is the standard metal used in nanoscale integrated circuits today. From the comparison it will be clear that an individual CNT is not competitive against copper. Instead, a dense or tightly packed array of nanotubes is essential for practical interconnect applications. We will conclude the discussion in this chapter by summarizing some of the practical challenges that must be resolved for CNT interconnect wires to be realized in an integrated circuit.

---

[1] An advanced treatment of electron scattering in solids can be found in M. Lundstrom, *Fundamentals of Carrier Transport*, 2nd edn. (Cambridge University Press, 2000).

7.2   Electron scattering and lattice vibrations

Electron transport in a conductor can be categorized into two different regimes. One is the ballistic regime, discussed in Chapter 6, and the other is the diffusive regime, where electrons experience repeated scattering during transport. In describing the transport of electrons in a metal wire or interconnect, one has to consider both the ballistic and diffusive regimes in order to obtain a comprehensive picture of interconnect resistance over the typical length scales involved. This section will elucidate on the sources of scattering in CNTs, and Section 7.3 will quantify the scattering in terms of the widely used metric known as the mean free path. In a sense, scattering provides an opportunity for electrons that have gained energy from an applied field to lose that energy and return to thermal equilibrium with the lattice.

Just as there are several methods by which a moving vehicle loses speed on a highway (road bumps, surface roughness, pot holes, stop signs, crowded lanes, stray animals, etc.), there are also several means by which an electron can experience a scattering event (and lose its speed) in a CNT including the following:

(i)   *Lattice vibrations*: The atoms in a solid are constantly vibrating about their mean position with increased displacement as a function of temperature. These vibrations are *quantized waves* and are otherwise known as *phonons*. Interaction of the mobile electrons and phonons is a major source of scattering.

(ii) *Lattice defects*: Defects or disorder in the regularity of the carbon atoms can lead to scattering of a mobile electron. Examples of typical defects include missing carbon atoms in the nanotube (known as vacancies) and the presence of extra atoms within the nanotube (interstitials).

(iii) *Electron–electron interactions*: The Coulombic repulsion between electrons results in deflection of the electrons. We will not consider this form of scattering in this text because at moderate temperatures (including room temperature) transport in CNTs has been described successfully within the independent electron formalism.

(iv) *Substrate roughness*: Nanotubes supported on a substrate can be disturbed by the roughness of the substrate surface, potentially resulting in the scattering of mobile electrons.

(v)   *Electron–substrate interactions*: Even on a substrate with an atomically smooth surface, electrons traveling in the nanotube can be deflected by the electric field originating from exposed charges present on the surface. This is particularly pertinent on polar substrates, such as silicon dioxide (SiO_2).

(vi) *Electron–superstrate interactions*: Similarly, exposed or unshielded charges in the superstrate[2] can deflect mobile electrons. Even in the absence of a superstrate, polar molecules in the surrounding ambient (e.g., H_2O) may adsorb onto the nanotube and lead to scattering of electrons.

---

[2] A superstrate is the counterpart of the substrate, and can represent an insulating film covering the nanotube.

Given the abundance of scattering possibilities, it is indeed quite remarkable that ballistic transport in CNTs is observable at room temperature for nanotube lengths of about 1 μm. To maintain a tractable discussion about scattering, it is useful to classify the scattering processes as either intrinsic or extrinsic scattering, where intrinsic scattering arises from processes inherent to the nanotube and extrinsic scattering refers to external disturbances that lead to electron scattering. We note that the first three enumerated processes are intrinsic, while the rest are extrinsic. Currently, the theory and understanding of extrinsic scattering pertaining to CNTs is in its infancy. Additionally, for metallic CNTs on high-quality substrates, a variety of experimental results suggest that electron scattering is limited by intrinsic scattering processes. For these reasons, we will focus on the intrinsic sources of scattering, mostly discussing lattice vibrations and, to a much lesser extent, CNT lattice defects.

We briefly mentioned earlier that lattice vibrations are due to the displacement of the carbon atoms from their nominal position. These displacements propagate along the length of the solid as a quantized wave or phonon with a finite velocity, albeit much slower than the electron velocity. In general, there are many types of vibration that can propagate in a solid. Each distinct type of vibration is known as a mode. For example, CNTs can have several dozen allowed modes. Some selected CNT lattice vibrations are shown in Figure 7.1, with their corresponding designated names. For the reader with little or no familiarity about phonons, it is most convenient to consider phonons at a high-level abstraction instead of focusing on the specific details of lattice vibrations. In this sense, phonons can be understood and described in much the same way as electrons, although without the charge. That is, phonons are particle-waves that propagate through the lattice characterized by an energy (or frequency)–wavevector dispersion relation. Additionally, we know that the number of electrons can be determined from the electron DOS and the probability of occupation (Fermi–Dirac distribution). Likewise, the number of phonons at any given temperature can also be determined from the phonon DOS and the corresponding probability of occupation, which is governed by the Bose–Einstein distribution.[3] While electrons are responsible for charge transport, phonons are mostly responsible for heat transport in CNTs.[4]

The phonon dispersion in CNTs is fairly complicated and mathematically involved. Fortunately, our interest here is to obtain a basic idea of the properties of phonons, and appreciate the important consequence that the interaction of electrons and phonons can lead to scattering which manifests electrically in the form of resistance. Figure 7.2 shows the phonon dispersion of a metallic CNT to give an

---

[3] Fermi–Dirac distribution applies to quantum particles with half-integer spin, such as electrons, while Bose–Einstein distribution applies to quantum particles with integer spin, such as phonons and photons.

[4] E. Pop, D. A. Mann, K. E. Goodson and H. Dai, Electrical and thermal transport in metallic single-wall carbon nanotubes on insulating substrates. *J. Appl. Phys.*, **101** (2007) 093710. E. Pop, D. Mann, J. Cao, K. Goodson and H. Dai, Nagative differential conductance and hot phonons in suspended nanotube molecular wires. *Phys. Rev. Lett.* **95** (2005) 155505.

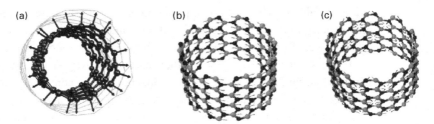

Fig. 7.1   Selected lattice vibrations of CNTs. The arrows indicate the direction of the displacement of the carbon atoms. (a) The radial breathing mode (RBM), which reflects the breathing of the nanotube,[5] leading to a time-periodic expansion/contraction of the diameter (courtesy of S. Reich *et al.*[6]). The twist (TW) mode, which is important for acoustic phonon scattering, and the zone-boundary A'_1 mode, which contributes to optical phonon scattering, are shown in (b) and (c) respectively. Courtesy of S. Roche, J. Jiang, L. E. F. Fon Torros and R. Saito, Charge transport in carbon nanotubes: quantum effects of electron–phonon couplings. *J. Phys. Condens. Matter*, **19** (2007) 183203. Copyright IOP Publishing Ltd (2007).

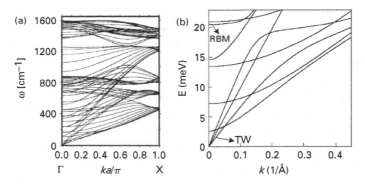

Fig. 7.2   The phonon dispersion of a (10, 10) metallic CNT. The full spectrum within the first half of the Brillouin zone is shown in (a) and the low-energy spectrum is shown in (b) highlighting some modes, including the TW and RBM modes. Each curve or branch corresponds to a distinct lattice vibration or mode. The modes that begin at the origin are acoustic modes, while the rest are optical modes. The Γ and X points are the high-symmetry points of the CNT's Brillouin zone. Note that the y-axis can be expressed interchangeably as either the spatial frequency w or energy E. (Adapted from R. Saito *et al.*[7]).

idea of what these dispersions look like. In solid-state physics, lattice vibrations are broadly classified as either acoustic modes leading to *acoustic phonon* propagation or optical modes leading to *optical phonon* propagation. The acoustic phonons have

---

[5] The RBM mode is an especially popular and routinely accessible optical mode for characterizing the diameter of nanotubes by Raman spectroscopy.

[6] S. Reich, C. Thomsen and J. Maultzsch, *Carbon Nanotubes: Basic Concepts and Physical Properties* (Wiley-VCH, 2004).

a vanishing energy or frequency in the small wavevector or long wavelength limit, and with a relatively high velocity can propagate sound waves (hence the use of the term acoustic) through the lattice. The optical phonons have a finite energy even at zero wavevector and can be excited by light or photons, which explains why they are labeled as optical. For readers interested in an advanced treatment of phonons in CNTs, two comprehensive expositions on the topic are quite insightful.[4,7]

It is expedient to consider a thought experiment to refine our intuition regarding electron scattering by phonons. In CNTs, electron transport is 1D; that is, electrons are either traveling to the right or to the left. Let us consider an electron traveling to the right, for example. This electron will necessarily have a positive velocity. In the ideal (ballistic) case, such an electron will travel through the CNT from the left contact to the right contact without experiencing any scattering. For the case where lattice vibrations cannot be ignored, the interaction of the electron with a phonon can lead to backscattering of the electron, resulting in reduced conductance. That is, its direction has been altered by the scattering event and it is now traveling to the left (with negative velocity). This is in contrast to bulk metals, where electrons can scatter in the three dimensions of space. Diagrams illustrating the most studied electron–phonon scattering events in CNTs are illustrated in Figure 7.3.

The backscattering of an electron by a phonon is governed by the laws of conservation of energy and momentum (or wavevector):

$$E_i = E_f \pm \hbar w_{ph}, \tag{7.1}$$

$$k_i = k_f \pm k_{ph}, \tag{7.2}$$

where E_i and E_f the initial and final energy of the electron respectively and k_i and k_f are the initial and final wavevector of the electron respectively. k_{ph} is the wavevector of the phonon, $\hbar$ is the reduced Planck's constant, and w_{ph} is the temporal frequency of the phonon. $\hbar w_{ph}$ is the energy of the phonon, with a positive sign indicating an absorbed phonon and a negative sign indicating an emitted phonon. In the theory of electron transport in metallic nanotubes, the locations of the phonons that are thought to be responsible for backscattering of electrons have been largely restricted as follows (with reference to Figure 7.2):

(i) *zone-center phonons* – these are acoustic or optical phonons with wavevectors close to the Γ-point of the phonon dispersion ($k \approx 0$);

(ii) *zone-boundary phonons* – these are acoustic or optical phonons with wavevectors close to the X-point of the phonon dispersion.

This restriction of the phonons along the high-symmetry points is mostly for the sake of theoretical simplicity and to advance intuitive understanding. Moreover, it

---

[7] R. Saito, G. Dresselhaus, and M. S. Dresselhaus by *Physical Properties of Carbon Nanotubes* (Imperial College Press, 1998).

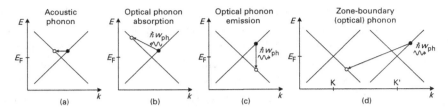

Fig. 7.3   Idealized diagrams illustrating electron–phonon scattering in metallic CNTs. (a) Acoustic phonon scattering. Optical phonon absorption and emission scattering processes are shown in (b) and (c) respectively. (d) Zone-boundary phonon scattering. (a) and (b) are intraband scattering processes; and (c) and (d) represent interband scattering.

spares us the monumental task of having to *accurately* derive (and study) the detail structure of the phonon dispersion for arbitrary wavevectors, which is needless to say rather complicated.[8] Ultimately, this restriction and the resulting theoretical simplicity are widely accepted because they are successful in explaining experimental observations of electron transport in CNTs even though phonons anywhere in the Billouin zone generally have a probability to scatter electrons.

New let us interpret acoustic and optical phonon scattering of electrons in metallic CNTs in the context of this restriction and with reference to Figure 7.2. With regard to the acoustic mode we will focus only on zone-center acoustic phonons, because so far it has not been necessary to invoke zone-boundary acoustic phonons in understanding experimental observations. In the vicinity of the zone-center, acoustic phonons have very small energies compared with electron energies;[9] as such, E_f is approximately equal to E_i in Eq. (7.1). Hence, acoustic phonon scattering can be considered to be approximately elastic and the scattered electron remains within the subband, as shown in Figure 7.3a. For optical phonons in the vicinity of the zone-center, there are a variety of phonon energies to consider. Therefore, for the general case, an electron scattered by an optical phonon will either gain energy (optical phonon absorption) or lose energy (optical phonon emission) during the scattering event, as illustrated in Figure 7.3b and Figure 7.3c respectively. Near the zone-boundary, optical phonons have significant energies and momentum. The wavevectors of these phonons are comparable to the distance (in k-space) between the K and K′ points of graphene's Brillouin zone. Therefore, for the conservation of momentum to be satisfied, electrons in a subband centered, say, at the K′-point will be scattered into another subband centered at the K-point, as shown in Figure 7.3d. To understand this conclusion, it is necessary to recall

[8] J. Jiang, A. Saito, G. G. Sammanidze *et al.*, Electron–phonon matrix elements in single-wall carbon nanotubes. *Phys. Rev. B*, **72** (2005) 235408.

[9] CNT electron emission experiments determined an absolute value of the Fermi energy to be approximately 1.72 eV. See P. Wu N. Y. Huang, S. Z. Deng, S. D. Liang, J. Chen and N. S. Yu, The influence of temperature and electric field on field emission energy distribution of an individual single-wall carbon nanotube. *Appl. Phys. Lett.*, **94** (2009) 263105.

from Chapter 4 that, with respect to the Brillouin zone of graphene, nanotubes have two conduction (or valence) bands that are degenerate at the Fermi energy. One conduction band is located at the K-point and the other is at the K′-point.

In brief, we can summarize that acoustic phonons lead to (approximately) elastic electron backscattering that is within the subband. On the other hand, optical phonons lead to inelastic electron backscattering that can be either intraband or interband.[10]

7.3   Electron mean free path

Considering lattice vibrations is a means to an end in the study of CNT interconnects. It is desirable for the end result to be a metric or set of metrics that captures and quantifies the scattering of electrons by phonons. At the end of the day, we can then conveniently include the scattering metric(s) in the formulation of the material resistance. For this purpose, several metrics have been defined in solid-state physics to quantify the scattering of electrons, including the electron *scattering time, lifetime*, and the *mean free path*. These metrics are average quantities and are meaningful only because electrons typically scatter repeatedly when traveling in a solid. Hence, one can define the average time (lifetime) it takes before an electron experiences a scattering event or, equivalently, the average length (mean free path) between scattering events. In semiconductor materials, it is the norm to convert the lifetime to another metric known as the mobility. In metals, including metallic CNTs, the mean free path is the more widely used currency because it is experimentally easier to infer its value.

An important observation from Figure 7.2 is that phonons are characterized by their energy, which is largely supplied thermally; hence, the number of phonons is a strong function of temperature. Coupled with the fact that electrons are also characterized by energy which can be supplied both thermally and electrically (from the applied voltage), we therefore expect intuitively that the scattering of electrons by phonons will generally be a function of both temperature and applied voltage or electric field. For example, as the nanotube temperature increases, the number of phonons increases substantially, resulting in enhanced interaction or scattering of electrons, which consequently leads to a higher resistance as a function of temperature. The bias conditions are often divided into the low-field and high-field conditions, with a transitional region in between. We will elaborate further on these different bias conditions as we proceed. As a prelude, at room temperature, acoustic phonon scattering is the dominant scattering mechanism under low-field conditions, whereas optical phonon scattering is dominant under high-field conditions.

---

[10] Sometimes these are referred to as intrasubband or intersubband scattering or, alternatively, intravalley or intervalley scattering.

Each scattering mechanism is characterized by its own mean free path. For example, the electron mean free path due to acoustic phonon scattering l_{ac}, and optical phonon scattering l_{op} can be separately extracted and identified under different experimental temperature and bias conditions. Likewise, the mean free path due to defect scattering l_d can be extracted typically at very low temperatures of a few kelvin when the number of phonons is negligible. The composite electrical parameter that includes all the different mean free paths is the effective mean free path $l_{m,eff}$. The effective mean free path is computed according to Matthiessen's rule, which is similar to the summing of resistances in parallel:

$$l_{m,eff} = \left(\frac{1}{l_{ac}} + \frac{1}{l_{op}} + \frac{1}{l_d} \right)^{-1}. \tag{7.3}$$

The optical phonon scattering mean free path consists of two terms characterizing absorption and emission processes:

$$\frac{1}{l_{op}} = \frac{1}{l_{op,abs}} + \frac{1}{l_{op,ems}}, \tag{7.4}$$

where $l_{op,abs}$ and $l_{op,ems}$ are the optical phonon absorption and emission mean free paths respectively. The key insight is that the effective mean free path is mostly determined by the shortest of all the mean free paths under the operating conditions. The question that now arises is: What are the functional dependencies and working expressions for all the separate mean free paths in order to compute $l_{m,eff}$? Let us broadly address the question first and then move on to determine the working expressions for each term in Eq. (7.3). As discussed previously, electron scattering is generally a function of temperature and bias potential. Additionally, since nanotubes come in different diameters, it is reasonable to expect a diameter dependence. It then follows by extension that in general

$$l_{m,eff} = f(T, \mathcal{E}, d_t), \tag{7.5}$$

where T is the temperature, ε is the applied field or bias ($\mathcal{E} = V/L$, V is voltage, and L is the CNT length), and d_t is the CNT diameter.

Electron scattering by lattice defects is mostly independent of temperature because lattice defects are determined by the quality of the CNT synthesis procedure and to first order can be considered to be invariable after synthesis. Additionally, lattice defects are relatively static in nature (relative compared with the very mobile electrons); hence, electron scattering is largely elastic, involving no loss of electron energy. This implies that it is not necessary to account for electron energy or bias; hence, scattering by lattice defects can be considered bias independent. However, we do expect a diameter dependence because nanotubes of different diameters might have different densities of lattice defects. Fortunately, the diameter dependence appears to be very weak. Purewall *et al.*,[11] have reported experimentally extracted values for l_d of $\sim 7–9\ \mu$m for metallic CNTs with diameters from $\sim 1–2.5$ nm. For simplicity, a suitable value for l_d can be taken as a constant, representative of electron scattering in high-quality metallic CNTs, and used for modeling and interpretation of experimental data.

Theoretical calculations have revealed that the (zone-center) acoustic phonon scattering in metallic CNTs is primarily due to the TW mode, with the resulting simplified expression for the acoustic phonon mean free path being

$$l_{ac} = \alpha_{ac}\frac{d_t}{T}. \tag{7.6}$$

Theoretical estimates for α_{ac} is approximately in the range $(400–565) \times 10^3$ K.[8,12] The physics of the expression for the acoustic phonon mean free path is as follows. The inverse dependence of l_{ac} on T comes about from the increased (decreased) number of acoustic phonons at higher (lower) temperatures, which increases (reduces) the probability of electron scattering. What is more, the linear dependence of l_{ac} on d_t is thought to arise from the reduced probability of backscattering due to the smaller *DOS per unit cell* (normalized differently compared with the standard DOS per unit length) as the CNT diameter increases.[13] Employing the lower estimate for α_{ac} (to be on the conservative side), $l_{ac} \sim 1.3–2.7\ \mu$m for CNTs with diameters of 1–2 nm at room temperature. Alternatively, the acoustic phonon mean free path can be expressed in a normalized manner to its value at room temperature ($T_{300} = 300$ K).

$$l_{ac}(T) = l_{ac,300}\frac{T_{300}}{T}, \tag{7.7}$$

where $l_{ac,300}$ is the room temperature acoustic phonon mean free path for the specific nanotube diameter of interest. The value of $l_{ac,300}$ can be assigned its theoretical estimate or extracted from measurements as a fitting parameter to model nanotube resistance.

Working expressions for the optical phonon mean free paths require a bit more care to derive, involving an interplay between interpretation of experimental observations and theoretical insights. The optical phonon scattering processes are strongly dependent on the number of optical phonons N_{op} present. In turn, the number of optical phonons, which is given by the Bose–Einstein distribution, is a

[11] M. Purewall, B. H. Hong, A. Ravi, B. Chandra, J. Hone, and P. Kim, Scaling of resistance and electron mean free path of single-walled carbon nanotubes. *Phys. Rev. Lett.*, **98** (2007) 186808.

[12] H. Suzuura and T. Ando, Phonons and electron–phonon scattering in carbon nanotubes. *Phys. Rev. B*, **65** (2002) 235412.

[13] J. Jiang, J. Dang, H. T. Yang and D. Y. Xing, Universal expression for localization length in metallic carbon nanotubes. *Phys. Rev. B*, **64** (2001) 845409.

strong function of temperature and the phonon energy:

$$N_{op}(T) = \frac{1}{\exp(E_{op}/k_B T) - 1},$$
(7.8)

where k_B is Boltzmann's constant. It has been discovered that there is a critical optical phonon energy, $E_{op} \sim 0.16$ eV, that is successful in explaining experimental measurements of electron transport in metallic CNTs.[14] This suggests that modes associated with this critical energy are predominantly responsible for electron scattering by optical phonons. In modeling of nanotube resistance, E_{op} is commonly employed as a fitting parameter with a value between ~ 0.15 and 0.2 eV.

We will largely be following the methodology of Pop *et al.*[4] in obtaining analytical relations for the optical phonon mean free paths in order to gain insight into their functional dependencies. The expression for the optical phonon absorption mean free path can be conveniently normalized to an appropriate value at room temperature:

$$l_{op,\,abs}(T) = l_{op,300} \frac{N_{op}(T_{300}) + 1}{N_{op}(T)},$$
(7.9)

where $l_{op,300}$ is the spontaneous optical phonon emission length at room temperature for a given nanotube diameter. To be clear, $l_{op,300}$ is the characteristic length scale for optical phonon emission by electrons whose energy already exceeds the critical phonon energy (meaning they can spontaneously emit optical phonons); that is, electrons with energy $> E_{op}$.[15] The actual optical phonon emission mean free path will be discussed shortly. Theoretical calculations indicate that $l_{op,300}$ is linear with nanotube diameter.[8] Invariably, the physics of $l_{op,abs}$ is similar to that of l_{ac} save for a stronger temperature dependence coming from the exponential Bose–Einstein distribution. In practice, $l_{op,300}$ is treated like a fitting parameter with values anywhere from $l_{op,300} \sim 10$ to 100 nm, frequently towards the lower end of the range. Evaluation of $N_{op}(T)$ for $T < 300$ K typically leads to values for $l_{op,abs} > 10\,\mu$m, which is much greater than l_{ac}. As a result, electron scattering via the optical phonon absorption process is sometimes neglected in simplified low-temperature models of CNT resistance.

For electron scattering via optical phonon emission, there are two processes to consider. One process involves an electron that has gained energy in excess of E_{op} from the electric field and subsequently emits an optical phonon. Another scattering process involves an electron that has absorbed an optical phonon and subsequently emits that optical phonon after traveling a certain distance. It follows that an average optical phonon emission mean free path is a Matthiessen sum of

[14] Z. Yao, C. Kane and C. Dekker, High-field electrical transport in single-wall carbon nanotubes. *Phys. Rev. Lett.*, **84** (2000) 2941–4.

[15] In CNTs, electrons with energies greater than the critical phonon energy $E_{op} \sim 0.16$ eV are sometimes referred to as hot electrons.

these two scattering processes:

$$l_{op,ems} = \left(\frac{1}{l_{ems,fld}} + \frac{1}{l_{ems,abs}} \right)^{-1}, \qquad (7.10)$$

where $l_{ems,fld}$ is the emission scattering length due to electrons that have gained sufficient energy from the electric field and $l_{ems,abs}$ is the emission scattering length due to electrons that have gained sufficient energy by absorbing an optical phonon. $l_{ems,fld}$ is given as

$$l_{ems,fld}(T) = \frac{E_{op}}{e|\mathcal{E}|} + \frac{N_{op}(T_{300}) + 1}{N_{op}(T) + 1} l_{op,300}, \qquad (7.11)$$

where e is the electric charge. The first (field-dependent) term is an estimate for the distance an electron has to travel in the electric field to acquire sufficient energy. This distance is computed by solving for the length from the two basic electrostatic relations, force $=$ energy/length and force $= e\mathcal{E}$. The latter temperature-dependent term in Eq. (7.11) is the average distance the electron travels after it has acquired the sufficient energy, ultimately resulting in the emission of an optical phonon. In the same sense, $l_{ems,abs}$ is a sum of the average length it takes for an electron to absorb an optical phonon plus the average length to emit the optical phonon afterwards:

$$l_{ems,abs}(T) = l_{op,abs}(T) + \frac{N_{op}(T_{300}) + 1}{N_{op}(T) + 1} l_{op,300}. \qquad (7.12)$$

Now is a great time to pause and reflect on the expressions for the mean free paths to arrive at some important insights. Defect scattering mean free path can approximately be taken as a constant value independent of temperature and bias. The expressions for the phonon mean free paths reveal a temperature dependence for all three scattering mechanisms (l_{ac}, $l_{op,abs}$, and $l_{op,ems}$), and a bias or field dependence for $l_{op,ems}$. While the focus has largely been on the temperature and field dependence of the phonon mean free paths, the reader should keep in mind that both $l_{ac,300}$ and $l_{op,300}$ have a theoretically predicted linear dependence on diameter which has recently been verified experimentally; see Liao et al.[16] It is notable that, in the low-field moderate temperature regime, the effective mean free path is approximately determined by l_{ac}, and this regime is of significant interest because it corresponds to room-temperature operation of nanotube interconnects. Since the low-field regime is of practical importance, it would be worthwhile to derive an expression for a maximum or threshold electric field which demarcates this regime. This could be achieved, for example, by defining the maximum electric field as that which satisfies the criteria $l_{ac} = 10 l_{op}$. However, such an expression

---

[16] A. Liao, Y. Zhao and E. Pop, Avalanche-induced current enhancement in semiconducting carbon nanotubes. *Phys. Rev. Lett.* **101** (2008) 256804.

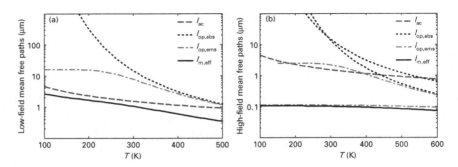

Fig. 7.4   Temperature dependence of the various electron mean free paths due to phonon scattering.
The only field-dependent term is the optical phonon emission mean free path.
(a) Low-field mean free paths ($\mathcal{E} = 10$ mV μm^{-1}). (b) High-field mean free paths
($\mathcal{E} = 2$ Vμm^{-1}). The parameters used for these plots are $E_{op} = 0.16$ eV,
$l_{ac,300} = 1.5$ μm, $l_{op,300} = 30$ nm, and l_d (not shown) is assigned a value of 10 μm.

is cumbersome owing to the lengthy terms in l_{op} and its complex dependence on
temperature. For analysis and measurements, it is convenient to have a numerical
estimate or range of what low-electric fields imply. In experimental investigations
of nanotubes, low electric fields are typically <10–20 mV μm.

Figure 7.4 offers a visual representation of the temperature dependence of the
electron mean free paths due to the various phonon scattering processes. At low
temperatures ($\lesssim$300 K) in the low-field limit, $l_{m,eff}$ is dominated by acoustic
phonon scattering with an approximately $1/T$ dependence. At room temperature and above, optical phonon scattering is no longer negligible, resulting in
an enhanced decline in the low-field effective mean free path with a stronger temperature dependence ($\sim 1/T^2$) above $\sim$300 K. The flat profile of $l_{op,ems}$ at low T
occurs because this is largely determined by the field-dependent first term in Eq.
(7.11). At higher T in the low-field limit, it is actually the optical phonon absorption process that confers a strong temperature dependence on $l_{op,ems}$. For the case
of high electric fields, the effective mean free path is largely determined by $l_{op,ems}$
at typical operating temperatures,[17] and its specific value is inversely proportional
to the applied bias. Elucidation of high fields in CNTs is presented Section 7.5.

7.4   Single-wall CNT low-field resistance model

The resistance of a material is the basic parameter that determines the suitabil-
ity of the material as an interconnect or wire. Ideally, the interconnect resistance

---

[17] Caution should be exercised when extracting the optical phonon mean free path or $l_{op,300}$ from
experimental data, especially under high fields, where it is most accessible. The challenge lies in
determining the temperature that corresponds to the extracted value, because often-times the CNT
temperature under high fields can be much higher than the ambient temperature.

should be vanishingly small or negligible compared with the resistance of the devices connected at the ends of the interconnect. In an era where mobile gadgets are ubiquitous and energy consumption is of paramount concern, a relatively small interconnect resistance offers many benefits, including the option to reduce interconnect dimensions for lower capacitance. Altogether, smaller resistance and lower capacitance are important in order to reduce power dissipation and increase battery life. We recall from Chapter 6 that the resistance of an ideal individual metallic nanotube connected to transparent contacts is the quantum resistance

$$R_q = \frac{h}{4e^2}, \tag{7.13}$$

where h is Planck's constant and the factor of 4 accounts for the total number of modes due to the spin and subband degeneracy. Numerically, $R_q \sim 6.45\,\text{k}\Omega$. The quantum resistance is only valid when electrons travel through the interconnect without scattering. That is, when the length l of the nanotube is much shorter than the electron (effective) mean free path. Hence, for arbitrary nanotube lengths greater than the mean free path, electrons will experience scattering resulting in increased resistance beyond R_q. Additionally, the contacts to the nanotube may not be fully transparent, leading to a parasitic or residual resistance R_{res}. Therefore, the total resistance of an arbitrary length of single-wall CNT of a certain diameter can be expressed in the Landauer form:

$$R(L, T, V) = R_{\text{res}} + R_q \left[1 + \frac{L}{l_{\text{m,eff}}(T, V)} \right] = R_c + R_q \frac{L}{l_{\text{m,eff}}(T, V)}. \tag{7.14}$$

R_c represents the total contact resistance ($R_c = R_{\text{res}} + R_q$) and the term in parentheses is the inverse of the Landauer transmission probability mentioned in Chapter 6 (see Eq. (6.11)). The latter term is the length-dependent nanotube resistance, as illustrated in Figure 7.5a. The universal profile of the length-dependent resistance is shown in Figure 7.5b. In general, the resistance profile can be separated into two length scales to facilitate understanding of the underlying physics. At short length scales ($L \gg l_{\text{m,eff}}$), the CNT approaches ballistic transport with $R \sim R_c$, roughly independent of length. For longer length scales ($L \ll l_{\text{m,eff}}$) the CNT exhibits diffusive transport characteristic of classical metals and the resistance per unit length is

$$\frac{R}{L} = \frac{R_q}{l_{\text{m,eff}}}, \quad L \gg l_{\text{m,eff}}, \tag{7.15}$$

and is roughly independent of the contact resistance. This is, of course, provided the contact resistance is reasonable, say within a factor of 10 of the quantum

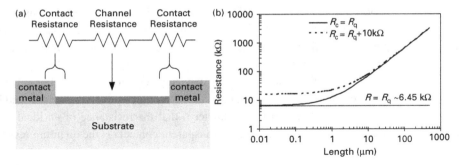

Fig. 7.5   (a) Illustration of a nanotube wire between two identical contacts. Each contact contributes a resistance of $R_c/2$, and the total CNT resistance is the sum of the contact and length-dependent channel resistances. (b) Example of the total CNT resistance profile for the case when $l_m = 1\ \mu$m. The solid curve is for an ideal R_c, while the dashed curve includes an additional 10 kΩ contact resistance. For reference, the CNT quantum resistance is shown.

resistance.[18] Eq. (7.15) is sometimes referred to as the 1D resistivity. For routine analysis, it is often convenient to have an order of magnitude estimate of the resistance. With this in mind, for a mean free path of 1 μm, which is realistic for high-quality CNTs operating in the low-field condition at room temperature, $R/L \sim 6.45\ \text{k}\Omega\,\mu\text{m}^{-1}$ at long length scales. At intermediate length scales, the specific curvilinear nature of the resistance profile has to be taken into account. For applications that require uniform and repeatable interconnect resistance, this intermediate length scale is the least robust, since it is sensitive to variations in both the contact resistance and the mean free path.

Eq. (7.14) is a general equation that describes the resistance of the nanotube as a function of length, temperature, and field or voltage bias. As such, it is applicable for low, intermediate, and high electric fields. Of special interest is the low-field regime where most practical interconnects operate. In this regime, $l_{m,eff}$ is largely determined by acoustic phonon scattering (with a small correction for optical phonon scattering), which yields a CNT resistance that is roughly independent of the bias voltage. For short length scales, when ballistic condition applies, this resistance is also independent of temperature, whereas at longer length scales $R \sim 1/T$ or $R \sim 1/T^2$, depending on the temperature range, as discussed in the previous section. A working expression for the low-field resistance R_{lf} is desirable, particularly under ambient conditions (in air, at room temperature). This working

---

[18] The full understanding of metal–nanotube contacts and contact resistance is a matter of research. In practice, palladium (Pd) is widely employed to provide nearly transparent contacts to metallic CNTs. This is partly because Pd has a high workfunction relatively close to that of CNTs; and, moreover, Pd is fairly insensitive to moisture. See D. Mann, A. Javey, J. Kong, Q. Wang and H. Dai, Ballistic transport in metallic nanotubes with reliable Pd ohmic contacts. *Nano Lett.*, **3** (2003) 1541–4.

expression can be written as

$$R_{\text{lf}}(L) \approx R_{\text{c}} + R_{\text{q}}L\left(\frac{1}{l_{\text{ac}}} + \frac{1}{l_{\text{d}}} + \frac{1}{l_{\text{op,lf}}}\right), \tag{7.16}$$

where $l_{\text{op,lf}}$ represents the small correction term due to optical phonon scattering at low fields. This correction term can be obtained from a zeroth-order Taylor series approximation for l_{op} (Eq. (7.4)):

$$l_{\text{op,lf}} \approx \frac{l_{\text{op,300}}(E_{\text{op}} + e|\mathcal{E}|l_{\text{op,300}})e^\Delta}{2E_{\text{op}} + e|\mathcal{E}|l_{\text{op,300}}e^\Delta}, \quad \Delta = \frac{E_{\text{op}}}{k_{\text{B}}T_{300}}, \tag{7.17}$$

which when substituted into Eq. (7.16) provides an expression for R_{lf} that is useful for rapid analysis in the vicinity of $T = 300$ K. An even cruder estimate for the low-field resistance suitable for back-of-the-envelope calculations[19] is simply to neglect the l_{d} and $l_{\text{op,lf}}$ terms in Eq. (7.16). For numerical computations, the complete expression for the effective mean free path (Eq. (7.3)) can of course be conveniently employed.

In practical interconnect applications, straight aligned arrays of metallic CNTs are extremely sought after because of the much smaller resistance per unit length that is achievable. We will discuss more about arrays of nanotubes in Section 7.10, particularly within the context of performance comparison with copper interconnects. The next section will examine the CNT resistance under high electric fields.

7.5 Single-wall CNT high-field resistance model and current density

Under high-field conditions (typically a few volts per micrometers) $l_{\text{m,eff}} \approx l_{\text{op,ems}}$ at practical operating temperatures, including hundreds of degrees above room temperature, resulting in a CNT resistance that is relatively independent of temperature (see Figure 7.4b) but linear with the field strength or voltage bias. A working expression for the field-dependent nanotube resistance at a constant temperature is derived by constructing an effective mean free path consisting of acoustic phonon scattering (important at low fields) and the field-dependent term in Eq. (7.11) for $l_{\text{ems,fld}}$, which is significant at high fields. That is:

$$R(L, V) = R_{\text{c}} + R_{\text{q}}\frac{L}{l_{\text{m,eff}}} \approx R_{\text{c}} + R_{\text{q}}\frac{L}{l_{\text{ac}}} + R_{\text{q}}L\frac{e|\mathcal{E}|}{E_{\text{op}}}. \tag{7.18}$$

---

[19] Back-of-the-envelope is a popular phrase among scientists/engineers and refers to the practice of using very simplified models to perform quick hand calculations and obtain immediate results. Historically, this used to be performed on the back of an envelope or a nearby random piece of paper. In today's digital lifestyles, there are many such *equivalent envelopes*, including a tablet PC.

For a nanotube of a given length, the resistance can be expressed explicitly in terms of the voltage bias:

$$R(V) \approx R_c + R_q \frac{L}{l_{ac}} + \frac{eR_q}{E_{op}}|V|, \tag{7.19}$$

which can be organized into bias-independent and dependent terms:

$$R(V) \approx R_o + \frac{|V|}{I_o}, \tag{7.20}$$

where $R_o = R_c + R_q L / l_{ac}$, $I_o \, (=E_{op}/eR_q)$ is commonly called the saturation current, for reasons that will be obvious shortly.[20] Let us calibrate our intuition to some typical numbers for the saturation current. For $E_{op} \sim 0.16$–0.2 eV, $I_o \sim 25$–31μA. It is called the saturation current because the current in the nanotube approaches I_o at high voltages.[21] To see this, let us examine the CNT current–voltage relation:

$$I(V) = \frac{V}{R(V)} = \frac{I_o V}{I_o R_o + |V|}. \tag{7.21}$$

At increasingly large voltages greater than $I_o R_o$, the current $I \rightarrow I_o$. Figure 7.6 is the measured large-bias current and resistance of an experimental nanotube device showing the current saturation and voltage-dependent resistance.

Equations (7.20) and (7.21), while successful in interpreting high-field experimental observations, are nonetheless fundamentally semi-empirical in nature. A complete physics-based model would, as a requirement, take into account the non-equilibrium temperature profile along the length of the nanotube. This non-equilibrium condition arises from hot electrons losing energy to the lattice via scattering (also termed Joule heating), thereby leading to temperatures at the center of the nanotube that can be hundreds of degrees higher than the lower temperature maintained at the contacts to the nanotube. From a modeling or parameter extraction point of view, the major consequence of a non-uniform temperature profile is that it becomes difficult to ascertain the effective temperature of an extracted mean free path parameter under high bias. For this purpose, a working physics-based electro-thermal model that accounts for the non-equilibrium temperature profile has been developed by Pop *et al.*[4] and the reader will find the model useful for high-field electron transport as an alternative to the semi-empirical models of Eqs. (7.20) and (7.21).

[20] For the reader interested in the history of E_{op}, it was from experimental observation of a saturation current in CNTs that the idea of the critical phonon energy came to life courtesy of Dekker and coworkers (see footnote 14). E_{op} is normally calculated from experimental extraction of I_o in nanotube I–V measurements under high bias.

[21] This simplified model for the field-dependent resistance and current saturation values is valid for long nanotubes ($L > l_{m,eff}$). For shorter nanotubes, transport is ballistic or quasi-ballistic, and the current saturates at larger values or may not saturate at all but instead grows linearly with voltage until nanotube breakdown.

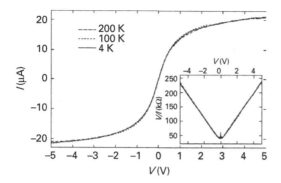

Fig. 7.6   The high-bias current of a 1 μm long metallic nanotube at different temperatures showing current saturation. The inset plots the voltage-dependent resistance. Adapted figure with permission from Yao *et al.*[14] Copyright (2000) by the American Physical Society.

An important metric for interconnects is the maximum current density. An estimate of the maximum current density in metallic nanotubes can be calculated by dividing the saturation current ($\sim$25–31μA) by the cross-sectional area of the nanotube. For metallic single-wall nanotubes of diameters $\sim$1–3 nm operating in ambient conditions, the maximum current density limited by burn-out (or oxidation in air) by Joule heating is of the order of $\sim$10^9 A cm^{-2}. To appreciate this value, it is worthwhile comparing with the maximum current density of copper, which is the interconnect metal in advanced very large-scale integrated (VLSI) technology. Experimental studies show that, depending on the cross-sectional profile, copper interconnects in VLSI technology have a maximum current density of the order of 10^7–10^8 A cm^{-2}.[22] In this regard, metallic nanotubes offer a factor of $\sim$10–100 improvement in the current density, and this is one of the reasons why CNTs are being explored as an interconnect material in advanced technologies.

Further increases in the electric field or voltage bias eventually lead to significant heating of the nanotube, which raises the maximum temperature along the nanotube until breakdown occurs, characterized by an abrupt decline of the current from its saturation value to zero at the breakdown voltage. In air, breakdown occurs by the process of oxidation of the carbon atoms in the nanotube by oxygen molecules, resulting in an irreversible loss of carbon atoms and permanent gaps in the physical structure of the nanotube. Estimates of the breakdown field are around 5 V μm^{-1} for long nanotubes ($L > l_{\mathrm{m,eff}}$) operating in ambient conditions.[4]

[22] This maximum current density is due to burn-out of the copper wires. See G. Schindler, G. Steinlesberger, M. Engelhardt and W. Steinhögl, "Electrical characterization of copper interconnects with end-of-roadmap feature sizes." *Solid-State Electron.*, **47** (2003) 1253–6.

7.6   Multi-wall CNT resistance model

Multi-wall CNTs with their multiple concentric shells that can contribute to charge transport are of significant interest as high-performance interconnects for future advanced technologies, where great emphasis is placed on obtaining the largest conductance in the smallest geometry. In Chapter 6, ballistic conduction in an ideal MWCNT was discussed; the ballistic quantum resistance including the participation of all the shells is

$$R_{q,mw} = \frac{h}{2e^2} \frac{1}{N_{ch}(V,T)}, \tag{7.22}$$

where N_{ch} is the total number of conducting channels. In general, N_{ch} is bias- and temperature-dependent. In this chapter, we will employ the average value of N_{ch} described in Chapter 6 and given by the following expression, which applies at room temperature under low-bias conditions:

$$N_{ch} \approx \left(\frac{d_{out} - d_{in}}{2\delta} + 1 \right) \left(a_1 \frac{d_{out} + d_{in}}{2} + a_2 \right), \tag{7.23}$$

where d_{in} and d_{out} are the inner and outer shell diameters respectively and $\delta \sim$ 0.34 nm is the shell-to-shell spacing. For $d_{in} \gtrsim 6$ nm, the fitting parameters are $a_1 \sim 0.107\,\text{nm}^{-1}$ and $a_2 \sim 0.088$. Equation (7.23) informs us that, to obtain the smallest ballistic resistance, the multi-wall nanotube should possess a large number of shells (first part of the expression) and a high average diameter (second part of the expression). This ballistic resistance applies when the length of the multi-wall nanotube is much less than its effective mean free path ($l_{m,eff}$). For multi-wall nanotubes of arbitrary length with contacts that may not be fully transparent,[23] the resistance at a given temperature and bias can be expressed in the Landauer form.

$$R(L) = R_{res} + \frac{h}{2e^2 N_{ch}} + \frac{h}{2e^2 N_{ch}} \frac{L}{l_{m,eff}} = R_c + R_{q,mw} \frac{L}{l_{m,eff}}. \tag{7.24}$$

R_c represents the total contact resistance ($R_c = R_{res} + R_{q,mw}$). The latter term is the length-dependent multi-wall nanotube resistance.

Unlike single-wall nanotubes, a comprehensive experimentally verified theory of electron scattering processes in multi-wall nanotubes and the effective mean free path is a matter of ongoing theoretical and experimental investigations. Specifically, electron scattering due to lattice vibrations, disorder in the arrangement of carbon atoms, and defects in the lattice are questions of primary concern. Preliminary theoretical and experimental studies have suggested electron mean free paths

---

[23] We continue to retain the assumption that all the shells are contacted for the purpose of obtaining insight regarding the maximum performance limits of multi-wall nanotubes. In reality, contacting all the shells has so far been a fairly rare event.

of the order of tens of micrometers for high-quality MWCNTs with outer diameters about 50–100 μm.[24] Perhaps for greatest practical relevance, $l_{m,eff}$ should be considered an experimentally determined statistical parameter for the diameter distribution of multi-wall nanotubes produced from a particular fabrication method.

Notwithstanding the lack of an experimentally verified comprehensive theory of electron scattering in multi-wall nanotubes, it is still possible to obtain further insights regarding the resistance of MWCNTs that will be especially valuable in assessing the performance of multi-wall nanotubes as a potential interconnect in advanced technologies. The desire to assess the performance of a candidate metal compared with other competing candidates for interconnect applications leads to the metric of resistivity (or, alternatively, conductivity). The resistivity metric is useful because it normalizes out the length and cross-sectional geometry, which facilitates comparison of the performance of different materials that might otherwise form in different shapes (e.g. square materials versus tubular materials). The resistivity ρ of a MWCNT is

$$\rho = R\frac{A}{L} = \frac{R_c A}{L} + \frac{R_{q,mw}A}{l_{m,eff}}, \tag{7.25}$$

where the MWCNT area is $A = \pi(d_{out}/2)^2$. Practical applications of multi-wall nanotubes as interconnects in advanced VLSI technology will undoubtedly need to have good contacts for low resistance. That is, it is reasonable to expect $R_c \approx R_{q,mw}$; hence:

$$\rho = R_{q,mw}A\frac{L + l_{m,eff}}{Ll_{m,eff}}, \tag{7.26}$$

which simplifies to

$$\rho = \frac{h\pi d_{out}^2}{8e^2 N_{ch}}\frac{L + l_{m,eff}}{Ll_{m,eff}}. \tag{7.27}$$

The conductivity σ is the reciprocal of the resistivity, $\sigma = 1/\rho$. In the ballistic regime, the resistivity is inversely proportional to length, and the dependence on length indicates it is not a material constant.[25] In the diffusive regime, the resistivity

---

[24] See J. Jiang et al.,[13] which theoretically explores the role of disorder in the backscattering of electrons. Also, see the experiment reported in H. J. Li, W. G. Li, J. J. Li, X. D. Bai and C. Z. Gu, "Multichannel ballistic transport in multiwall carbon nanotubes." *Phys. Rev. Lett.*, **95** (2005) 086601, which investigated an MWCNT with contacts to all (or almost all) the shells and suggests a mean free path of ~25 μm at room temperature.

[25] Material constants imply no dependence on length, height, and width of the material. While the resistivity is not a material constant in the ballistic regime, the ballistic resistance itself is a material constant.

is independent of length and becomes a material constant similar to conventional metals. A more explicit expression for the resistivity is achievable for MWCNTs with more than 10 shells. For this case, the factor of unity in Eq. (7.23) can be ignored and N_{ch} is

$$N_{ch} \approx \frac{d_{out}[a_1 d_{out}(1 - \beta_r^2) + 2a_2(1 - \beta_r)]}{4\delta}, \quad 0 < \beta_r < 1, \tag{7.28}$$

where β_r is the ratio between the inner and outer diameters according to the relation $d_{in} = \beta_r d_{out}$, and is considered to be of the order of $\frac{1}{2}$. Therefore, the simplified resistivity becomes

$$\rho \approx \frac{h}{2e^2} \frac{\pi d_{out}\delta}{a_1 d_{out}(1 - \beta_r^2) + 2a_2(1 - \beta_r)} \frac{L + l_{m,eff}}{L l_{m,eff}}. \tag{7.29}$$

In Section 7.10 the value of the resistivity/conductivity as a metric becomes clear when the performance of MWCNTs as interconnects is benchmarked against copper interconnects.

7.7   Transmission line interconnect model

The complete model of a CNT as an interconnect necessarily includes the quantum capacitance and kinetic inductance in addition to the electrostatic capacitance and magnetic inductance. Our attention will be focused on the low-energy regime where most interconnects operate. For single-wall metallic nanotubes the quantum capacitance per unit length C_q and kinetic inductance per unit length L_K are respectively (from Chapter 6)

$$C_q = \frac{8e^2}{hv_F} = \frac{2}{R_q v_F} = 310 \text{ aF}\,\mu\text{m}^{-1} \tag{7.30}$$

and

$$L_k = \frac{h}{8e^2 v_F} = \frac{R_q}{2v_F} \equiv 3.2 \text{ nH}\,\mu\text{m}^{-1} \tag{7.31}$$

where v_F is the Fermi velocity ($\sim 10^6$ m s^{-1})[26] and $R_q \sim 6.45$ kΩ. Similarly, the quantum capacitance and kinetic inductance of multi-wall nanotubes are functionally identical to those of single-wall nanotubes, but modified by the number of

---

[26] Some earlier papers on CNTs employed $v_F \sim 8 \times 10^5$ m s^{-1}, which resulted in $C_q \sim 400$ aF μm^{-1} and $L_k \sim 4$ nH μm^{-1}. Subsequent experimental measurements have since estimated the Fermi velocity $\sim 10^6$ m s^{-1}.

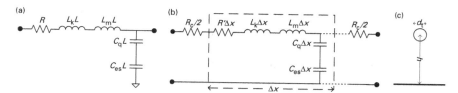

Fig. 7.7 (a) Lumped model of a nanotube wire or interconnect. L is the length of the CNT. (b) Transmission line model of a nanotube interconnect. The transmission line can be viewed as a cascade of sections with an incremental length of Δx. (c) Cross-section of a transmission line consisting of a CNT over a ground plane.

channels that the MWCNTs offer:[27]

$$C_q = \frac{N_{ch}}{R_q v_F}, \tag{7.32}$$

$$L_k = \frac{R_q}{v_F N_{ch}}. \tag{7.33}$$

As discussed in Chapter 6, the quantum capacitance manifests electrically as a capacitance in series with the electrostatic capacitance, and the kinetic inductance is likewise in series with the magnetic inductance and the wire resistance, as shown in Figure 7.7a. In the circuit theory of electromagnetic waves, Figure 7.7a is considered a lumped model in the sense that the resistances, capacitances, and inductances can be represented as discrete individual circuit elements. The lumped model is valid when the wire or interconnect length is much less than the wavelength of the (voltage or current) signal of interest, because in this regime the wave or oscillatory nature of the signal is not apparent. For much longer wire lengths (or alternatively much higher signal frequencies), the signal can oscillate in amplitude and phase along the length of the wire. Therefore, the wire must be considered a distributed system, as shown in Figure 7.7b. This distributed model is commonly called a transmission line model.[28] The transmission line model is necessary to account properly for the signal delay through the wire, the signal attenuation as it travels down the wire, and the dispersion of spectral components of the signal. As a general rule of thumb, the distributed model becomes increasingly relevant for

[27] The same symbols C_q and L_k will be used to denote the quantum capacitance and kinetic inductance of both single-wall and multi-wall nanotubes because the analysis methods apply equally to both families of nanotubes.

[28] The transmission line model was developed by Oliver Heaviside to describe communication through long cables (circa 1885). The life of Mr. Heaviside is an inspirational study of the perseverance to master any topic. Unemployed and living with his parents, he managed to teach himself physics and mathematics and went on to develop vector calculus, which enabled him to condense Maxwell's original (20) equations into the four differential equations known today. He also invented the Heaviside step function, patented the coaxial cable, and coined many accepted terms, including conductance, impedance, and inductance.

interconnect lengths greater than a tenth of the signal wavelength (see Figure 7.8 for likely values of the wavelength for single-wall CNT transmission lines).

In the transmission line model explored in this chapter, ohmic loss in the insulating substrate is not considered a valid approximation for low-loss insulators such as quartz and silicon dioxide (SiO_2). In brief, the circuit elements in Figure 7.7 are

$R_c/2 \equiv$ input and output contact resistance
$R' \equiv$ resistance per unit length (excluding R_c); R is the CNT resistance including R_c; $R' = (R - R_c)/L$
$L_k \equiv$ kinetic inductance per unit length
$L_m \equiv$ magnetic inductance per unit length
$C_q \equiv$ quantum capacitance per unit length
$C_{es} \equiv$ electrostatic capacitance per unit length

It is worthwhile evaluating the typical values of the kinetic and magnetic inductances to determine their relative contribution to the total inductance. The magnetic inductance of a wire over a ground plane (see Figure 7.7c) is,[29]

$$L_m = \frac{\mu_o}{2\pi} \cosh^{-1}\left(\frac{2h}{d_t}\right), \tag{7.34}$$

where μ_o is the free-space permeability applicable to typical insulating substrates separating the CNT from the ground plane. The inverse hyperbolic cosine for the most part has a logarithmic profile and, as such, it is not particularly sensitive to large variations in h/d_t. For single-wall nanotubes of diameters from 1 to 3 nm and typical insulator thickness of 10 nm to 1 μm, the magnetic inductance is within a factor of two of $L_m \sim 1\,\mathrm{pH}\,\mu\mathrm{m}^{-1}$, which is about three orders of magnitude less than $L_k \sim 3.2\,\mathrm{nH}\,\mu\mathrm{m}^{-1}$ of single-wall CNTs. For multi-wall nanotubes of diameters from 10 to 100 nm and the same range of insulator thickness, the magnetic inductance is within a factor of four of $L_m \sim 0.4\,\mathrm{pH}\,\mu\mathrm{m}^{-1}$. If we consider a multi-wall nanotube with inner and outer diameters of 10 nm and 100 nm respectively, as representative of the largest MWCNT, its kinetic inductance is $\sim 8\,\mathrm{pH}\,\mu\mathrm{m}^{-1}$ ($N_{ch} \sim 797$ at room temperature), which is about an order of magnitude greater than its magnetic inductance. Therefore, we can conclude that, for both single-wall and multi-wall nanotubes of reasonable diameters on insulating substrates of typical thickness, the magnetic inductance is negligible compared with the kinetic inductance. Hence, we can neglect L_m from here onwards.

Likewise, comparison of the typical values of the quantum and electrostatic capacitances provides insight as to their relative importance. The electrostatic

---

[29] For a review of electromagnetics and transmission line theory see S. Ramo, J. R. Whinnery and T. Van Duzer, *Fields and Waves in Communication Electronics* (Wiley, 1994).

capacitance of a wire over a ground plane is

$$C_{es} = \frac{2\pi \varepsilon_0 \varepsilon_r}{\cosh^{-1}(2h/d_t)},$$ (7.35)

where ε_0 is the free-space permittivity, and ε_r is the dielectric constant of the insulating substrate. For single-wall nanotubes of diameters from 1 to 3 nm and typical insulator thickness (e.g. SiO_2) of 10 nm to 1 μm, the electrostatic capacitance is within a factor of two of $C_{es} \sim 50\,aF\,μm^{-1}$ which is of the same order of magnitude as the quantum capacitance of single-wall CNTs ($\sim 310\,aF\,μm^{-1}$), especially true with the use of higher dielectric constant substrates. For multi-wall nanotubes, C_{es} and C_q are also comparable, particularly for thinner MWCNTs. As a result, it is reasonable to conclude that both capacitances are significant. The total capacitance per unit length is

$$C_{tot} = \frac{C_q C_{es}}{C_q + C_{es}}.$$ (7.36)

The total capacitance can also be expressed to explicitly include some form of electron–electron interactions. The (Luttinger liquid) theory of electron–electron interactions in 1D conductors which is applicable to CNTs reveals that the fundamental excitations are collective charge waves and spin waves which propagate at different velocities.[30] An important concept that comes out of that theory is captured by a dimensionless parameter g and essentially quantifies the strength of the electron–electron interactions. One method of computing g is via a capacitive relation that includes both C_q and C_{es} and in a sense embodies electron interactions vis-à-vis the screened electrostatic Coulomb interaction. In this light, the interaction parameter is defined as,[31]

$$g = \sqrt{\frac{C_{tot}}{C_q}} = \sqrt{\frac{C_{es}}{C_{es} + C_q}},$$ (7.37)

where g is always less than unity for interacting electrons. A value of g close to unity signifies a weakly interacting system ($C_{es} \ll C_q, C_{tot} \rightarrow C_q$) while a value closer to zero signifies a strongly interacting system ($C_{es} \gg C_q, C_{tot} \rightarrow C_{es}$). For strictly non-interacting systems, $g = 1$.

[30] In 1D conductors, charge density and spin-density waves propagate at different velocities and this is known as spin–charge separation. Spin density propagates with a velocity $\sim v_F$ while charge density propagates at an enhanced velocity $= v_F/g$. For a theoretical discussion, see F. D. M. Haldane, "Luttinger liquid theory" of one-dimensional quantum fluids: I. Properties of the Luttinger model and their extension to the general 1D interacting spinless Fermi gas. *J. Phys. C: Solid State Phys.*, **14** (1981) 2585–2609.

[31] C. Kane, L. Balents and M. P. A. Fisher, Coulomb interactions and mesoscopic effects in carbon nanotubes. *Phys. Rev. Lett.*, **79** (1997) 5086–9.

The stage has now been set to discuss the properties of a nanotube transmission line. Principally, a transmission line is characterized by two basic parameters: (i) the characteristic impedance Z_o which informs us of the conditions for optimum signal transport into and out of the line and (ii) the propagation constant γ, a parameter that contains information regarding the attenuation and phase or dispersion of the signal as it travels along the line. From transmission line theory,[29] the characteristic impedance and propagation constant are

$$Z_o = \sqrt{\frac{R' + jwL_k}{jwC_{tot}}} = \sqrt{\frac{L_k}{C_{tot}}\left(1 + \frac{R'}{jwL}\right)}, \tag{7.38}$$

$$\gamma = \sqrt{(R' + jwL_k)jwC_{tot}} = jw\sqrt{L_kC_{tot}}\sqrt{1 + \frac{R'}{jwL}} = \alpha + j\beta, \tag{7.39}$$

where w is the signal frequency.[32] The real and imaginary components of γ are the attenuation constant α and the phase constant β respectively. The wavelength λ of signals propagating through the line is defined as

$$\lambda = \frac{2\pi}{\beta}. \tag{7.40}$$

A plot of the wavelength for a single-wall CNT transmission line is shown in Figure 7.8 from 1 GHz to 1 THz with $R' = 6.45$ kΩ μm^{-1} (valid for high-quality CNTs at room temperature). An estimate of the wavelength at some key frequencies is useful to keep in mind. At 1 GHz and 10 GHz, the wavelength is within a factor of two of $\lambda \sim 150$ μm and $\lambda \sim 45$ μm respectively, for C_{es} between $0.1C_q$ and $10C_q$. For multi-wall nanotubes, a typical value of R' is not commonly encountered, making an estimate of the wavelength not as useful. Nonetheless, some values of the wavelength of MWCNT transmission lines will be estimated in the next section.

It is notable that the characteristic impedance and propagation constant are generally functions of frequency. Indeed, we can define a critical frequency w_z to demarcate two regimes of operation:

$$w_z = \frac{R'}{L_k}. \tag{7.41}$$

In the technical language of signals and systems theory, this critical frequency is known as the zero frequency w_z.[33] The significance of w_z becomes clear when the

---

[32] For convenience, both symbols w and f will be known as frequency. They are interrelated via the expression $w = 2\pi f$

[33] More precisely $-jw = R'/L_k = w_z$ is the root or zero of $y = R' + jwL_k$.

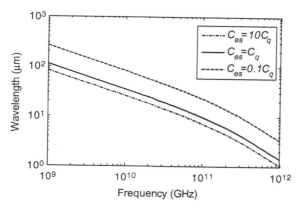

Fig. 7.8 Wavelength of single-wall nanotube transmission lines at room temperature.

properties of the transmission line are examined for the following conditions

$$w \ll w_z \text{ (or } wL_k \ll R') \to \text{lossy transmission lines}$$
$$w \gg w_z \text{ (or } wL_k \gg R') \to \text{lossless transmission lines.}$$

Lossless (no signal attenuation) and lossy (attenuates signal) nanotube transmission lines are elucidated in the subsequent sections. For the moment, it is useful to compute a numerical estimate for w_z in order to have some perspective regarding the transition from lossy to lossless CNT transmission lines. For single-wall CNTs, $R' \sim R_q/l_{m,eff}$, and with $l_{m,eff} \sim 1\mu m$ at room temperature, it follows that

$$f_z = \frac{1}{2\pi}\left(\frac{R_q}{l_{m,eff}}\frac{2v_F}{R_q}\right),$$
$$f_z = \frac{v_F}{\pi l_{m,eff}}, \tag{7.42}$$

which for high-quality single-wall nanotubes at room temperature with $l_{m,eff} \sim 1\,\mu m$ yields $f_z \sim 318\,\text{GHz}$. Equation (7.42) also applies to MWCNTs and for $l_{m,eff}$ of the order of $10\,\mu m$, f_z is of the order of $31.8\,\text{GHz}$. We can therefore conclude that single-wall nanotubes at room temperature will be operating as lossy transmission lines at gigahertz frequencies and gradually transition to a lossless transmission line at sub-terahertz frequencies. A lossless multi-wall nanotube transmission line is in principle possible at gigahertz frequencies at room temperature if the mean free path is large enough. Theoretically, the mean free paths of MWCNTs are diameter dependent; hence, a relatively thick MWCNT would be most attractive for lossless gigahertz transmission lines at room temperature. In the next two sections the focus will be on elucidating the properties of lossless and lossy nanotube transmission lines respectively.

7.8   Lossless CNT transmission line model

Lossless transmission lines are the most desirable type of transmission line because an incident signal can propagate through the line without attenuation and distortion, as will be seen shortly. Lossless propagation in CNTs are in principle observable at rather high frequencies satisfying the condition $w \gg w_z$ or $wL_k \gg R'$. In this regime, the characteristic impedance is

$$Z_0 = \sqrt{\frac{L_k}{C_{\text{tot}}}}, \tag{7.43}$$

a purely real frequency-independent constant. This indicates that the incident voltage and current signals along the line will be in phase. The characteristic impedance can be expressed conveniently in terms of the interaction parameter g:

$$Z_0 = \sqrt{\frac{L_k}{C_q}\left(\frac{C_{\text{es}} + C_q}{C_{\text{es}}}\right)} = \frac{1}{g}\sqrt{\frac{L_k}{C_q}}. \tag{7.44}$$

Notably, electron interactions enhance the characteristic impedance. For single-wall and multi-wall nanotubes, $Z_0 \sim R_q/(2g)$ and $Z_0 \sim R_q/(N_{\text{ch}}g)$ respectively. One of the experimental challenges in characterizing single-wall nanotubes is that Z_0 is of the order of 6.5 kΩ (say g is $\sim$0.5), which is very difficult to measure with standard high-frequency instrumentation that commonly have impedances of 50 Ω. The challenge facing multi-wall nanotubes is to contact all or most of the shells in order to obtain a significant number of channels which will aid in reducing its characteristic impedance to within a reasonable factor of 50 Ω.

The propagation constant of a lossless transmission line is

$$\gamma = jw\sqrt{L_k C_{\text{tot}}} = jw\sqrt{L_k C_q\left[\frac{C_{\text{es}}}{C_{\text{es}} + C_q}\right]} = gjw\sqrt{L_k C_q}. \tag{7.45}$$

γ is purely imaginary, entailing that there is no attenuation of the signal as it propagates through the transmission line. The phase constant is given by

$$\beta = gw\sqrt{L_k C_q}. \tag{7.46}$$

This leads directly to the definition of the group velocity v_g

$$v_g = \frac{\partial w}{\partial \beta} = \frac{1}{g\sqrt{L_k C_q}} = \frac{v_F}{g} \tag{7.47}$$

The group velocity is essentially the wave or signal velocity, or more precisely the velocity at which the envelope of the wave propagates through the transmission line. The constant v_g of a nanotube transmission line is indicative of dispersionless signal propagation, which is highly desirable in the communication of signals.[34] Notably, the interaction parameter enhances the signal velocity. The signal wavelength is $\lambda = v_g/(gf)$, and as an example, for MWCNTs operating at 50 GHz with $g \sim 0.5$, $\lambda \sim 40\,\mu m$.

7.9 Lossy CNT transmission line model

At frequencies much less than the zero frequency ($w \ll w_z$ or $wL_k \ll R'$), CNT transmission lines will be lossy due to ohmic loss in R', leading to a potentially substantial attenuation of the incident signal between the input and output of the line. In this regime, the transmission line is in essence a distributed RC network, and the characteristic impedance is given by

$$Z_0 = \sqrt{\frac{R'}{jwC_{tot}}} = \sqrt{\frac{R'}{2wC_{tot}}}(1-j) = \frac{1}{g}\sqrt{\frac{R'}{2wC_q}}e^{-j\pi/4}, \qquad (7.48)$$

where the complex identity $\pm\sqrt{j} = \pm(1+j)/\sqrt{2}$ has been employed to separate the real and imaginary components of Z_0. The negative imaginary term implies that the line is partially capacitive with a frequency-dependent amplitude and a phase difference of 45° between incident voltage and current waves propagating or diffusing through the line. The propagation constant is given by

$$\gamma = \sqrt{jwR'C_{tot}} = \sqrt{\frac{wR'C_{tot}}{2}}(1+j) = g\sqrt{\frac{wR'C_q}{2}}(1+j). \qquad (7.49)$$

The real and imaginary parts of γ yield the attenuation and phase constant respectively:

$$\alpha = \sqrt{\frac{wR'C_{tot}}{2}}, \qquad (7.50)$$

$$\beta = \sqrt{\frac{wR'C_{tot}}{2}}. \qquad (7.51)$$

---

[34] A constant group velocity means that all the frequency components of an incident signal will travel through the transmission line at the same speed; hence, the signal will not be distorted. On the other hand, a frequency-dependent v_g corresponds to frequency components of a signal traveling at different speeds through the line. Consequently, the frequency components will arrive at slightly different times at the output of the line, resulting in signal dispersion, an (undesirable) distortion of the signal.

For a transmission line of length L and an attenuation constant α, the incident signal will be attenuated by a factor of $e^{-\alpha L}$ at the output of the line. This exponential attenuation is often unacceptable in long transmission lines and, as a result, amplifiers (commonly called repeaters for this practice) are periodically inserted into the line to compensate for the attenuation.

The group velocity is

$$v_g = \frac{\partial w}{\partial \beta} = 2\sqrt{\frac{2w}{R'C_{tot}}}. \tag{7.52}$$

The frequency-dependent group velocity leads to dispersion of the input signal. The signal dispersion is further compounded by the fact that the attenuation constant is also frequency dependent. To minimize signal attenuation and dispersion, the line length should be kept to a minimum and the electrostatic capacitance should be much smaller than the quantum capacitance such that $C_{tot} \sim C_{es}$ to reduce the sensitivity of v_g to small changes in w.

7.10   Performance comparison of CNTs and copper interconnects

In modern advanced integrated circuits, copper is currently the interconnect of choice because of its higher electrical and thermal conductivity compared with other pure bulk metals (with the exception of silver). However, owing to aggressive scaling of dimensions in VLSI technology, the effective conductivity of copper and bulk metals has continuously degraded, in part due to the surface scattering of electrons and grain boundary scattering. Additionally, as the structural size of wires decreases in integrated circuits, electromigration phenomena[35] lead to significant reliability issues which cannot be ignored in high current-density interconnects. The decreased conductivity and reliability issues, both of which have been exacerbated at nanoscale dimensions, are now a primary concern in the development of future VLSI technologies. As a result, there is a need for superior next-generation interconnect materials suitable for VLSI. This brings us to metallic CNTs. Compared with copper, the relatively long mean free paths of nanotubes coupled with their higher maximum current density and resistance to electromigration (owing to the strong C–C covalent bond) makes them a promising candidate as interconnects in future integrated circuits. This section surveys reported performance assessment of single-wall and multi-wall nanotubes compared with copper for nanoscale integrated circuits.

---

[35] Electromigration refers to the transport or diffusion of ionized atoms from their original positions in a conductor, which can lead to electrical short and/or open circuits, a major circuit failure and reliability issue.

Figure 7.9 compares the conductivity $\sigma = L/(RA)$ of bundles of single-wall and multi-wall nanotubes of various diameters to copper interconnects in the low-bias regime at room temperature (reproduced from the model of Naeemi and Meindl).[36] In order to obtain numerical results, $d_{in}/d_{out} = \beta_r = 0.5$ is assumed in the MWCNT modeling. The single-wall CNTs have 1 nm diameter with a 1 μm mean free path. They are arranged to be densely packed bundles (one metallic nanotube per 3 nm^2 cross-sectional area) corresponding to one conducting CNT for every three nanotubes in order to evaluate the ultimate performance benefits that random chirality distribution of single-wall CNTs can potentially offer for interconnect applications. Let us take a moment to discuss the conductivity comparison. Copper has a flat conductivity at these length scales since its resistance increases with length, which normalizes out the length in the conductivity calculation. On the other hand, CNTs have a linear conductivity below their mean free path because the resistance is ballistic, and a constant conductivity above their mean free path due to the diffusive (length-dependent) resistance. We can conclude that CNTs, particularly thick multi-wall nanotubes, offer the largest conductivity at long length scales (so called semi-global and global interconnects), while copper is superior at much shorter length scales (in the local interconnect regime). However, this is an incomplete picture for practical applications where a device such as a transistor is driving the wire. In such cases, for local interconnects, the wire resistance is typically insignificant compared with the device resistance. Hence, for local wires the metric of most relevance is the wire capacitance, and CNTs continue to be attractive, since their capacitance will be smaller or comparable to that of copper owing to their smaller size.

Another important metric for interconnects is the RC time delay. The modeling of the wire capacitance to compute RC delays is fairly sensitive to the modeling assumptions and particular circuit application with appreciable differences in the models that have been reported in the literature. For this reason, it has been difficult to obtain a consensus qualitative picture of the RC delays of CNTs compared with copper. Needless to say, several workers have shown that densely packed arrays of single-wall nanotubes and thick multi-wall nanotubes are potentially several times faster than copper at long length scales.[37]

It is important to emphasize again that it is the qualitative insights offered by the CNT versus copper comparisons that are of greatest value. The precise quantitative benefits are undoubtedly sensitive to the model particulars and parameters,

[36] A. Naeeni and J. D. Maindl, Performance modelling for action nanotube intercorrects. In *Carbon Nanotube Electronics*, ed. A. Javey and J. Kong (Springer, 2009) pp. 163–90.
[37] See K.-H. Koo, H. Cho, P. Kapur and K. C. Saraswat, Performance comparison between carbon nanotubes, optical, and Cu for future high-performance on-chip interconnect applications. *IEEE Trans. Electron Devices*, **54** (2007) 3206–15; and A. Naeemi and J. D. Meindl, Performance modeling for single- and multiwall carbon nanotubes as signal and power interconnects in gigascale systems. *IEEE Trans. Electron Devices*, **55** (2008) 2574–82.

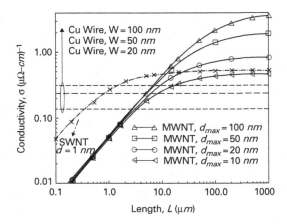

Fig. 7.9   Low-bias conductivity of densely packed single-wall nanotube bundles and multi-wall nanotubes of various diameters benchmarked to the conductivity of nanoscale copper wires. Reprinted from Naeemi and Meindl[36], copyright (2009), Springer Science+Business Media. With kind permission of Springer Science+Business Media.

and different parameters may produce appreciably different numerical results. The results from the performance benchmarking suggest that thick multi-wall nanotubes (with high d_{out}/d_{in} ratio) are the most attractive, both in terms of conductivity and delay metrics at long length scales. It is likely that a hybrid system or alloy of copper and nanotubes will offer sufficient performance for future interconnects.[38]

7.11   Summary

This chapter has explored the electrical properties of metallic single-wall and multi-wall nanotubes suitable for interconnect analysis at both short and long length scales. The electrical properties include mean free path, resistance, capacitance, and inductance. From an application point of view, the electrical properties are essential for developing physical models to aid in performance assessment of CNTs for future integrated circuits. At the moment, performance benchmarking of (single-wall and multi-wall) CNTs against copper indicates that nanotubes offer superior conductivity at length scales greater than their mean free path. However, there are several grand challenges that must be overcome before CNTs can be realized in integrated circuits. The grand challenges include the following:

(i) *Fabricating dense arrays or bundles of metallic CNTs*: Much progress has recently been made in synthesizing dense horizontally aligned planar arrays

---

[38] In addition to the overall improved conductivity a Cu–CNT alloy may potentially offer (over pure copper), the electromigration resistance and reliability of copper has been reported to be significantly enhanced (by about a factor of 5) by adding even a small amount of CNTs (~5 % by weight). See Y. Chai and P. C. H. Chan, High electromigration-resistant copper/carbon nanotube composite for interconnect application. *IEEE IEDM Tech. Digest*, (2008) 607–10.

of random chirality, single-wall nanotubes with average densities of about 5–10 CNTs/μm. However, for single-wall CNTs to fulfill their potential for superior interconnect performance compared with copper, we desire dense arrays of purely metallic nanotubes with a pitch between nanotubes to be of the order of nanometers (5 nm pitch corresponds to 200 metallic CNTs/μm). This is certainly not trivial to achieve and would require sustained research effort with the hope that a breakthrough will come forth in a reasonable time. Additionally, the dense arrays would have to be scalable to three dimensions. Likewise, synthesizing dense arrays or bundles of multi-wall nanotubes is an ongoing challenge.

(ii) *Chirality or diameter control*: Even if a technique to manufacture dense arrays of single-wall nanotubes were developed, it would be highly desirable if the chirality of the nanotubes in the array could be controlled in a routine manner. Ideally, all the nanotubes would be metallic of the same or similar diameters for interconnect applications. As an alternative option, semiconducting nanotubes that may be present in the array can be appropriately doped to become metallic. The doping techniques would have to be air stable and VLSI compatible to be of any practical use. Furthermore, the doped semiconducting CNTs should offer mean free paths comparable to intrinsic metallic nanotubes in order to maximize electrical performance.

(iii) *Low-temperature VLSI*: While it is possible to conceive several possible methods of integrating CNTs onto an integrated circuit, the least disruptive VLSI-compatible method is likely to be at low temperatures characteristic of contemporary back-end interconnect fabrication procedures.

(iv) *Contact resistance*: Contact resistance remains a persistent issue. It is not uncommon to see contact resistance variations of an order of magnitude from metallic single-wall nanotubes produced at the same time on the same substrate with identical contact metals. Similarly, making transparent contacts to all the shells of a multi-wall CNT has proved challenging with the use of standard contact fabrication techniques. It is encouraging that even with the relatively poor contact resistance routinely achievable at present, multi-wall nanotube interconnects have been shown to carry gigahertz signals.[39] To transport much faster signals will undoubtedly require advanced techniques to obtain transparent contacts to all the shells in order to obtain lower resistances.

While these collections of challenges are very complex, they are certainly not insurmountable. The sustained progress on nanotube research across many disciplines is the key to overcoming these issues.

---

[39] G. Close, S. Yasuda, B. C. Paul, S. Fujita, and H.-S. P. Wong, Measurement of subnanosecond delay through multiwall carbon-nanotube local interconnects in a CMOS integrated circuit. *IEEE Trans. Electron Devices*, **56** (2009) 43–9.

7.12   Problem set

All the problems are intended to exercise and refine analytical techniques while providing important insights. If a particular problem is not clear, the reader is encouraged to re-study the appropriate sections. Invariably, some problems will involve making reasonable approximations or assumptions beyond what is already stated in the specific question in order to obtain a final answer. The reader should not consider this frustrating, because this is obviously how problems are solved in the real world.

7.1. CNT mean free path comparison.
   Compare the low-field mean free path of metallic nanotubes at room temperature with that of standard bulk metals such as gold, silver, copper, aluminum, and platinum.

   (a) Relatively, how large is the CNT mean free path? Give an order of magnitude estimate.
   (b) Why is the CNT mean free path greater than that of standard bulk metals?

7.2. "High-quality" versus "low-quality" CNTs.
   Interconnect applications require high-quality metallic nanotubes in order to reap the performance benefits of CNTs compared with conventional metals such as copper or aluminum. However, the current synthesis techniques have yet to be perfected to routinely produce high-quality nanotubes. High-quality nanotubes in this context imply CNTs with a large defect-scattering electron mean free path l_d. Let us study the performance degradation that comes about from low-quality metallic nanotubes by taking its $l_d \sim 1\,\mu$m. The key idea here is to get a quantitative feel about the impact of low-quality metallic nanotubes. The room-temperature mean free path due to acoustic phonon scattering can be taken to be $1\,\mu$m.

   (a) What is the effective mean free path and resistance of the low-quality nanotube compared with the high-quality nanotubes at low fields at room temperature? Summarize the final results in percentage terms.
   (b) What is the additional power dissipation of the low-quality CNT interconnect if it has to carry $1\,\mu$A of current?
   (c) In the high-field limit at room temperature, estimate the impact of the low-quality nanotube on the effective mean free path.
   (d) Redo part (a) for vanishingly low temperatures.

7.3. Statistical description of the electrical properties of CNT wires.
   Today's nanotube synthesis techniques typically produce CNTs with a dispersion in diameter and mixed character (some semiconducting and some metallic). If we could wake up the synthesis genie, our first request would be

to ask for a method to produce all metallic nanotubes with identical diameters for interconnect applications.

Unfortunately, we have yet to see the synthesis genie and for now we have to deal with what we have. As such, a theory of nanotube resistance for today's growth methods must be statistical in nature for greatest relevance. For simplicity, assume that somehow we stumbled upon a method to produce only metallic nanotubes with a diameter dispersion (the next best thing compared with our heart's desire). Consider a Gaussian diameter dispersion, for diameters ranging from $d_t \sim d_o$ to $3d_o$, and centered halfway.

(a) For an interconnect that contains a large array of these nanotubes, what is the average mean free path and resistance of the CNT interconnect for low fields? For $d_o \sim 0.5$ nm, provide numerical estimates. Assume room temperature operation, $l_d \sim 10\,\mu\text{m}$ and $\alpha_{ac} \sim 400 \times 10^3$ K.
(b) What is the standard deviation of the mean free path and resistance?
(c) By good fortune we managed to wake up the synthesis genie, who granted us our desires of an array of nanotubes with each CNT having a 1 nm diameter. Compare the mean free path and resistance of this array with those of part (a).

7.4. Temperature dependance of CNT mean free path.
It is worthwhile to be mathematically fluent in routinely analyzing the temperature dependence of the CNT mean free path and resistance.

(a) For this purpose, plot the low-field mean free path of l_{ac}, l_d, $l_{op,abs}$, $l_{op,ems}$, and $l_{m,eff}$ for temperatures from 100 to 500 K. Use the values, $\mathcal{E} = 20\,\text{mV}\,\mu\text{m}^{-1}$, $E_{op} = 0.16\,\text{eV}$, $l_{ac,300} = 1.6\,\mu\text{m}$, $l_{op,300} = 15\,\text{nm}$, and $l_d = 50\,\mu\text{m}$.
(b) Show that, in the low-field limit, the effective mean free path is proportional to $\sim 1/T$ below room temperature and $\sim 1/T^2$ above room temperature. What is responsible for the inverse power of 2 dependence at higher temperatures compared with lower temperatures?

7.5. Single-wall versus multi-wall nanotubes.
Yes, nanotubes indeed compete with each other. A case in point is for interconnect applications. The pressing question is: Which type of nanotube provides the best performance? To address this question we explore an example model of both types of nanotube and restrict our attention to the low-field resistivity for the diffusive case at room temperature. For single-wall CNTs, we consider a dense square array of identical nanotubes ($d_t = 1$ nm, array cross-section has a height h), with a vertical and horizontal pitch. Also, the single-wall CNT has an acoustic phonon mean free path $\sim 1\,\mu\text{m}$. For multi-wall CNTs, $d_{in} = 6$ nm, and $d_{out} = h$.

(a) Derive the resistivity (classical resistivity in units of resistance-length) for both types of nanotube. Plot their resistivities on the same graph for $h = 20$ nm and 100 nm and for lengths from 100 nm to 100 μm. Assume the effective mean free path of the MWCNT is ~ 10 μm and diameter dependent to first order. Which type of nanotube offers the lowest resistivity? Also, determine which type of nanotube offers the lowest capacitance.

(b) It was noted in the text that $l_{m,eff}$ of multi-wall nanotubes is currently a matter of further research. As such, it is presently challenging to make a fair comparison involving multi-wall nanotubes in general, since we will have to assume some value for the mean free path. Obviously, very high values might be overly optimistic and likewise very low values might lead to negatively biased conclusions. Perhaps a compromise is to consider a range of values for the mean free path to get a more complete comparative picture. For this reason, let us restrict our attention to the diffusive limit, which makes it easier to perform the subsequent analysis. What is the expression for the resistivity of the MWCNT in this limit for an arbitrary $l_{m,eff}$?

(c) Plot the resistivity of both the single-wall and multi-wall nanotubes in the diffusive limit as a function of the multi-wall $l_{m,eff}$ ranging from 1 to 100 μm. From this plot, what can you conclude about the resistance of MWCNTs compared with single-wall CNTs for interconnect applications? Extend your qualitative conclusions to include comparisons with copper interconnects of the same cross-section.

8   Carbon nanotube field-effect transistors

> *Innovation is everything. When you are on the forefront, you can see what the next innovation needs to be.*
>
> Robert Noyce (co-inventor of integrated circuits and co-founder of Intel)

8.1   Introduction

In analogy to a water pipe that allows the guided flow of water, a transistor is an electronic device that allows for the guided flow of electrons with the key innovation being the influence of a gate that controls the amount of flowing electrons (the gate is similar in concept to a valve controlling the amount of water). The most popular flavor of the transistor is the *field-effect transistor* (FET), which came to reality in 1960[1] and forms the cornerstone of modern electronics that has revolutionized computing, communications, automation, and healthcare and fosters today's digital lifestyles. In part due to the continuous miniaturization or *scaling* of the transistor dimensions,[2] silicon (Si) has evolved to be the de facto semiconductor for making transistors that enable smaller, faster, cheaper, and more power-efficient integrated circuits (also called *chips*) for an extensive variety of applications. However, transistor scaling and the resulting performance enhancement cannot continue forever owing to both physical and technical reasons. Obviously, the transistor cannot be reduced to a size of zero length for example, and this imposes a physical limit to the miniaturization of devices. Fortunately, we have yet to reach this physical limit. At present, the more pressing issues are technical in nature: relating to the challenges of fabricating small transistors and, in addition, the significance at short size scales

---

[1] Interestingly, the FET was patented in 1926 by physicist Julius Lilienfeld. However, it took almost 35 years before a practical FET could be realized, partly because of the presence of surface states that usurped the influence of the gate. This issue was finally resolved in 1960 (at Bell Labs) by passivating the (silicon) semiconductor surface with an oxide (SiO_2). This brings to life the words of the anonymous writer, "Persistence is the twin sister of excellence. One is a matter of quality (innovation), the other, a matter of time (realization)."

[2] The great benefits of the scaling that has driven semiconductor technology over the past four decades was brought to light in the celebrated work of Robert Dennard and coworkers. See R. H. Dennard, F. H. Gaensslen, H.-N. Yu, V. L. Ridout, E. Bassous and A. R. LeBlanc, Design of ion-implanted MOSFET's with very small physical dimensions. *IEEE J. Solid-State Circuits*, **9** (1974) 25–68. Previously, Dennard had invented the DRAM memory cell.

of some otherwise undesirable device phenomena which are collectively referred to as short-channel or small-dimension effects.[3]

Largely due to the scaling challenges facing silicon transistors, there are substantial efforts to explore promising nanomaterials as candidates for future transistors with an implicit requirement that these nanomaterials not only produce transistors that scale gracefully (i.e. not sensitive to short-channel effects) at the nanoscale, but also offer higher performance at comparable length scales. This was the historic motivation for researching carbon nanotube field-effect transistors (CNFETs), where the semiconducting material is a (semiconducting) CNT. The experimental observation of ballistic electron transport over relatively long lengths in CNTs plus their higher current density and projected faster speed certainly make a compelling case for CNFETs as exploratory transistors. At the same time, CNTs, still being relatively new on the scene, deserve to be studied for their own sake as well, for a variety of reasons, including a necessary need to learn new paradigms about what nature offers in one dimension and the potential for new applications beyond the reach of conventional semiconductors. Hence, a comprehensive theory of the device physics of CNFETs is clearly of broad significance and interest that transcends the historical motivation.

The discussion begins with a survey of typical CNFET device geometries and the basic operating principles. Afterwards, an expression for the all-important CNT surface potential is developed. The middle sections of the chapter elucidate the theory of the device physics of CNFETs and comparisons with experimental results are examined. The chapter concludes by shining light on the paradigm shift between a 1D FET and its 2D counterpart. Much of the content from the previous chapters will be useful to the reader, with the lion's share of the content courtesy of Chapters 6 and 7. A basic understanding of silicon transistor device physics, while not required for understanding many of the ideas in this chapter, will nonetheless only help the reader in gaining broader and deeper insights. For example, basic device physics concepts such as energy band diagrams, workfunction, and flatband voltage will be a part of the narrative without any elementary introduction. The reader will find supplementary readings convenient to fill in any information gap.[3,4]

8.2   Survey of CNFET device geometries

Carbon nanotube FETs can be realized in several geometrical configurations. The most common device geometries are shown in Figure 8.1. The device geometries

3 See Y. Taur and T. H. Ning, *Fundamentals of Modern VLSI Devices*, 2nd edn (Cambridge University Press, 2008).
4 There are many outstanding texts, including R. F. Pierret, *Semiconductor Device Fundamentals* (Addison-Wesley, 1996); and B. G. Streetman and S. Banerjee, *Solid-State Electronic Devices*, 6th edn (Prentice Hall, 2005).

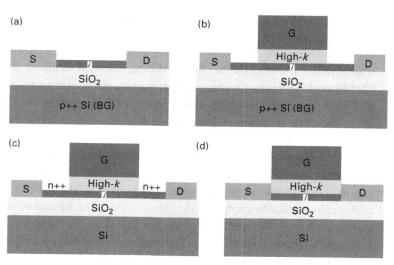

Fig. 8.1 Cross-sectional illustrations of four common CNFET device geometries. (a) A back-gated CNFET with the heavily doped Si substrate serving as the gate (p++ refers to heavy hole doping). (b) CNFET featuring both a top-gate and a back-gate. The top-gate is used to control the channel and the back-gate is used to (electrostatically) dope the CNT extension regions to achieve a low-resistance path to the S/D contacts. (c) A CNFET with chemically doped nanotube extension regions. (d) A self-aligned CNFET. Note that "i" represents an intrinsic semiconducting CNT.

all have certain familiar features, including an intrinsic (undoped) semiconducting nanotube serving as the channel, source (S) and drain (D) metallic contacts, a gate (G) electrode appearing in the form of a top-gate or a back-gate (BG), and an oxide or high-k dielectric,[5] to insulate the gate from the CNT. The operating dynamics of all the CNFET devices are very similar. The gate controls the amount of charge in the channel via a vertical electric field that induces either electrons or holes in the nanotube, while a horizontal electric field between the contacts provides the force that drives the charges from one contact to the other, resulting in an electric current. A top-gate is often preferred over a back-gate to provide localized and greater control of the nanotube channel, as well as to afford individual gate control in a multi-transistor circuit.

Figure 8.1a illustrates the most basic CNFET device geometry, where a heavily doped substrate serves as the gate. The CNFET device in Figure 8.1b has the unique distinction of featuring both a top-gate and a back-gate. To avoid any ambiguity, the actual semiconducting channel is the section of the nanotube directly underneath the top-gate. In this configuration, the top-gate is the main gate used to control the charge density in the nanotube channel, while the back-gate is employed to convert

[5] High-k dielectric is a fondly used jargon in the field of semiconductor physics. In a nutshell, it refers to dielectrics that possess a relative dielectric constant (k or ε_r) several times greater than that of SiO_2 ($k = 3.9$).

the semiconducting nanotube extension regions into quasi-metallic conductors in order to provide a low-resistance path between the channel and the metal contacts. This conversion is achieved by applying a sufficiently high back-gate voltage which creates a vertical electric field and induces excess electrons or holes in the CNT extension regions, a technique otherwise known as *electrostatic doping*. The CNFET device in Figure 8.1c is similar to that of Figure 8.1b, with chemical doping of the CNT extension regions instead of electrostatic doping. Figure 8.1d is an illustration of a *self-aligned* top-gated CNFET. It is referred to as self-aligned because, during fabrication, the gate metal is deposited first and subsequently used to align the source/drain metals, resulting in zero horizontal gap (or no extension regions) between the gate and the source/drain contacts. The self-alignment can be achieved in practice by exploiting the native oxide,[6] which forms on the surface of many metals when exposed to air or oxygen. For example, aluminum easily forms a native aluminum oxide (Al_2O_3). This native oxide around the gate metal is the enabling technique in obtaining self-alignment while providing electrical isolation between the gate and source/drain metals. Even though the CNFET devices shown in Figure 8.1 are supported by an SiO_2/Si substrate, it is entirely possible to employ a different substrate (for example, quartz or flexible substrates). The choice of silicon substrates is in part due to the advantages and convenience of leveraging the mainstream fabrication infrastructure that has been developed around bulk silicon. Similarly, the types of free carrier in the CNFET devices can be either holes (p++) or electrons (n++). A silicon substrate serving as a back-gate is often doped with holes because a lower substrate resistance can be achieved.

In general, charge transport in a CNFET can be categorized into four regimes regardless of the specific device geometry. The four transport regimes are identified in Table 8.1. They are determined by the length of the nanotube compared with their mean free path l_m and by the type of contact the nanotube makes with the source/drain metals. For example, an ohmic-contact ballistic CNFET refers to the case where charge injection from the S/D contacts into the CNT and vice versa is ohmic in nature (i.e. a simple resistance), and the charges travel through the nanotube channel without experiencing scattering. Conversely, a Schottky-barrier diffusive CNFET refers to a device where charges experience a Schottky barrier at the S/D nanotube junction, and charge transport through the nanotube channel suffers from repeated scattering. Charge can refer equally to electrons and/or holes. If electrons are the majority charge carriers, then the CNFET is an n-type transistor; and if holes are the majority carriers, then the CNFET is a p-type transistor. Quite intriguingly (and often undesired), it is also fairly easy to observe the so-called *ambipolar* behavior in CNFETs, where electrons are the majority

---

[6] A thin oxide, of the order of 1 nm, readily forms on the surface of many metals through the process of oxidation in the presence of air or oxygen. This natural thin oxide is what is called the native oxide.

Table 8.1. The four transport regimes of CNFETs

Metal–CNT contact	$L < l_{\mathrm{m}}$	$L > l_{\mathrm{m}}$
Ohmic	Ohmic-contact ballistic CNFET	Ohmic-contact diffusive CNFET
Schottky barrier	Schottky-barrier ballistic CNFET	Schottky-barrier diffusive CNFET

charges for positive gate voltages and holes are the majority charges for negative gate voltages.[7] This is in part due to their low bandgap.

We must point out that the four transport regimes are categorized with respect to the dominant charge carrier. This subtle point becomes important when considering an ohmic-contacted CNFET, because it is possible for the majority carrier (say electrons, for example) to experience an ohmic-contact at the metal–nanotube junction while holes simultaneously experience a Schottky barrier. An example that visually explains this distinction will be seen in Section 8.4. On the other hand, symmetric ambipolar CNFETs are always Schottky-barrier devices for (band diagram) symmetry reasons that will become clear in Section 8.8.

Theoretically, the contact type (ohmic or Schottky) is determined from the work-function difference between the contact metal and the CNT. In practice, so far p-type CNFETs are easier to encounter than n-type CNFETs for reasons related to the S/D–nanotube contact chemistry and physics. In short, for many of the conventional metals used as contacts, the Fermi energies of the metals more often than not align closer to the valence band of nanotubes, thereby favoring hole transport over electron transport. Even some metals whose Fermi levels are supposed to align closer to the conduction band are found to produce predominantly p-type devices, either due to Fermi-level pinning or chemical processes at the contact, such as oxidation of the metal at the metal–nanotube interface.

8.3 Surface potential

The surface potential φ_{s} is the local semiconductor electrostatic potential that accounts for the excess charge induced in the semiconductor due to an external electric field. It is the vital electronic property of all semiconductors that allows us to control its charge density to some degree. In a sense, the ability to control the surface potential (and by extension the charge density) is the defining characteristic of all semiconductors. A graphical description of the nanotube surface potential is shown in Figure 8.2. Note that the surface potential is intimately tied to the equilibrium Fermi energy E_{F}, since it accounts for the shift of the subbands with

---

[7] Conventional silicon FETs are designed to be unipolar. For example, electrons provide current in an n-type FET for positive gate voltages, and at negative gate voltages the FET is OFF (negligible current). And vice versa for p-type FETs.

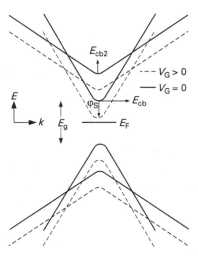

Fig. 8.2   A simplified band structure including the first two subbands of semiconducting nanotubes. E_{cb2} is the energy minimum for the second subband. The effect of V_G is to shift the bands up ($V_G < 0$) or down ($V_G > 0$) with respect to the equilibrium Fermi energy, resulting in a negative or positive φ_s respectively.

respect to the equilibrium E_F.[8] The polarity of the surface potential can be chosen arbitrarily. We define the surface potential as positive if the bands move down and negative if the bands move up with respect to their equilibrium position. This is a convenient convention because it bestows the polarity of the gate voltage on the surface potential. Furthermore, the polarity of φ_s instantly reveals the majority carrier in the nanotube:

$$\varphi_s > 0, \quad \text{n-type CNT}$$

$$\varphi_s = 0, \quad \text{intrinsic CNT}$$

$$\varphi_s < 0, \quad \text{p-type CNT.}$$

The surface potential is determined from the capacitive coupling of the gate to the nanotube, as discussed in Chapter 6:

$$\varphi_s = \frac{C_{ox}}{C_{ox} + C_q(\varphi_s)} V_G = T_{gc} V_G, \tag{8.1}$$

[8] Indeed, it is not uncommon for some authors to use the concept of the chemical potential in a virtually identical manner to surface potential. In that context, a positive V_G moves E_F up with respect to a static band, which is the reverse of the φ_s (see Figure 8.2), where the bands move relative to E_F. The departure of E_F from its equilibrium value is the chemical potential. We use φ_s to be consistent with conventional semiconductor transistor physics. In any case, the CNFET device physics depend only on the energy difference between the subbands and E_F, and both approaches lead to the same results.

where V_G is the gate voltage, C_q is the CNT quantum capacitance per unit length, and C_{ox} is the electrostatic capacitance per unit length between the nanotube and the gate metal, otherwise known as the oxide capacitance.[9] T_{gc} is the gate coupling transfer function. Equation (8.1) is a transcendental equation because the quantum capacitance has an exponential dependence on φ_s (see Chapter 6). Hence, φ_s has to be computed numerically by self-consistent iteration in general. However, for the purpose of obtaining insight into the nanotube device physics, a simple analytical expression for the surface potential would be most welcome. Moreover, an analytical expression would also be especially valuable in enabling compact modeling of CNFET devices and CNFET device/circuit design.

To aid in deducing an approximate analytical form for φ_s, it is worthwhile investigating the quantum ($C_q \ll C_{ox}$) and semiclassical (C_q comparable to C_{ox}) capacitance limits seperately. At sufficiently low gate voltages there is relatively smaller charge induced in the nanotube; hence, the CNFET will be in the *quantum capacitance limit* with $\varphi_s \approx 1$ and a unit slope ($T_{gc} = 1$). At higher gate voltages, both capacitances are comparable (*semiclassical limit*), and T_{gc} becomes sub-linear and less than unity. Mathematically, this is equivalent to

$$\varphi_s \approx \begin{cases} V_G, & C_q \ll C_{ox} \\ T_{gc}V_G, & C_q > C_{ox} \end{cases}. \tag{8.2}$$

Based on these insights, we can develop a phenomenological expression for the gate coupling function given by the relation

$$T_{gc}(V_G) = 1 - \left(1 - T_{gc,sc}(V_G)\right) H \left(1 - T_{gc,sc}(V_G)\right), \tag{8.3}$$

where $H(\cdot)$ is the Heaviside step function and $T_{gc,sc}(V_G)$ is an as-yet unknown gate coupling function that should capture the sub-linear behavior in the semiclassical limit. Observe that for $T_{gc,sc}(V_G) = 1$, $T_{gc} = 1$, corresponding to the quantum capacitance limit. The step function is simply employed as a convenient mathematical tool to merge the solutions in the quantum and semiclassical limits of operation. The challenge now lies in determining $T_{gc,sc}(V_G)$. We recall from Chapter 6 that the series combination of the quantum and oxide capacitances is approximately linear with V_G, at least over a restricted range. In that case, perhaps a first-order Taylor series expansion might suffice. Specifically, we desire the first-order series approximation of

$$T_{gc,sc} = \frac{C_{ox}}{C_{ox} + C_q(\varphi_s)}. \tag{8.4}$$

---

[9] For CNT devices with both a top and bottom gate, C_{ox} is expected to be the top-gate oxide capacitance, because the bottom-gate oxide is often much thicker (by design) with lower dielectric constant and, hence, has a relatively negligible influence on φ_s. In general, a complete model of such dual-gate devices should include the influence of both the top-gate and bottom-gate voltages.

We will now explore a mathematical brute force technique exploiting the Taylor series method in order to develop an analytical expression for the surface potential that will capture the functional dependence on C_{ox}, V_G, temperature T, and the CNT diameter or bandgap E_g. Let us proceed by first focusing on electrons in n-type devices (i.e., $V_G > 0$), and subsequently the results will be extended to holes in p-type devices. From Chapter 6, the electron quantum capacitance of a semiconducting CNT (including only the first subband)[10] is

$$C_q = \frac{2e^2 N_o \, e^{x_n} \left[1 + A \, e^{(\alpha x_n + \beta x_n^2)}(1 - \alpha - 2\beta x_n)\right]}{k_B T \left[1 + A \, e^{(\alpha x_n + \beta x_n^2)}\right]^2}, \qquad x_n = \frac{2e\varphi_s - E_g}{2k_B T}, \quad (8.5)$$

where A, α, and β are empirical constants in the analytical CNT carrier density theory determined to be 0.63, 0.88, and 2.41×10^{-3} respectively for temperatures around 300 ± 100 K (see Eq. (5.66)). N_o is the effective DOS:

$$N_o = \frac{8}{a} \frac{E_g + k_B T}{E_g + 4\gamma} \sqrt{\frac{k_B T}{3\pi E_g}}, \qquad (8.6)$$

where a is the graphene Bravais lattice constant ($\sim$2.46 Å), γ is the nearest neighbor overlap energy ($\sim$3.1 eV), and k_B is Boltzmann's constant. To make Eq. (8.5) significantly more manageable, we set $\alpha = 1$ and $\beta = 0$, since their empirical values are close to unity and zero respectively. These approximations are most welcome for hand-analysis of the subsequent calculations and generally result in negligible error in the low-energy regime. Of course, for a compact model using computational techniques, the approximations are unnecessary. With these approximations, C_q simplifies to

$$C_q \approx \frac{2e^2 N_o \, e^{\frac{2e\varphi_s - E_g}{2k_B T}}}{k_B T \left(1 + A \, e^{\frac{2e\varphi_s - E_g}{2k_B T}}\right)^2}. \qquad (8.7)$$

The first-order Taylor series approximation of Eq. (8.4) about $\varphi_s = mE_g$ results in the following expression for the surface potential (in the semiclassical limit):[11]

$$\varphi_s = (c_0 - c_1 \varphi_s)V_G, \qquad (8.8)$$

---

[10] Low-energy transport will be the primary focus of our exploration, and this is partly justified because this regime offers the highest performance nanotube transistors. Moreover, the low operating voltages in nanoscale technologies virtually guarantee that the low-energy regime is the most relevant. In this regime, electrons in the first subband dominate the electrostatic properties.

[11] The first-order Taylor series expansion of $C_{ox}/(C_{ox} + C_q) = 1/[1 + (C_q/C_{ox})]$ yields the binomial approximation $1/(1 + x) \sim 1 - x$, which is increasingly accurate as x becomes much smaller than unity.

where c_0 and c_1 are the Taylor series coefficients given by the lengthy expressions

$$c_0 = \frac{C_{ox}X[C_{ox}(k_BT)^2X^3 + e^2N_oZ]}{[C_{ox}k_BTX^2 + e^2N_oe^{(2m+1)E_g/2k_BT}]^2} \qquad (8.9)$$

where

$$X = e^{E_g/2k_BT} + Ae^{mE_g/k_BT}$$

$$Z = Ae^{E_g(4m+1)/2k_BT}(k_BT - mE_g) + e^{E_g(m+1)/k_BT}(k_BT + mE_g)$$

$$c_1 = \frac{e^2C_{ox}N_o\,e^{(2m+1)E_g/2k_BT}\left(A^2\,e^{2mE_g/k_BT} - e^{E_g/k_BT}\right)}{\left[C_{ox}k_BT\left(e^{E_g/2k_BT} + A\,e^{mE_g/k_BT}\right)^2 + e^2\,N_o\,e^{(2m+1)E_g/2k_BT}\right]^2}. \qquad (8.10)$$

The useful property of these cumbersome expressions is that they explicitly capture the dependence on C_{ox}, bandgap, and temperature, albeit in a somewhat complex manner. Ideally, if C_q was linearly symmetric about $e\varphi_s = 0.5E_{cb}$ then a natural choice for m corresponding to that point would be $\sim 0.25/e$. However, C_q is generally not symmetric, as such, a value of $m \sim (0.27/e) - (0.33/e)$ is found to give the best agreement with numerical self-consistent computation. Finally, solving for φ_s in Eq. (8.8) yields the analytical surface potential in the semiclassical limit.

$$\varphi_s \approx \frac{V_G}{sV_G + t}, \qquad (8.11)$$

with s and t related to the Taylor series coefficients by

$$s = \frac{c_1}{c_0}, \quad t = \frac{1}{c_0}. \qquad (8.12)$$

A similar analysis for holes ($V_G < 0$) produces the same expression for the surface potential with the exception that V_G in the denominator of Eq. (8.11) is replaced with $|V_G|$. Employing Eq. (8.11), we arrive at a final algebraic relation for the surface potential reflecting the influence of the gate in the quantum and semiclassical limits:

$$\varphi_s \approx V_G - V_G\left(1 - \frac{1}{s|V_G| + t}\right)H\left(1 - \frac{1}{s|V_G| + t}\right). \qquad (8.13)$$

Figure 8.3 shows the surface potential computed numerically and analytically for an example two-terminal top-gated CNT device. One terminal is the gate electrode and the other terminal is the source and drain electrodes, which have been shorted together (standard device connection for CNT capacitance measurements). The s

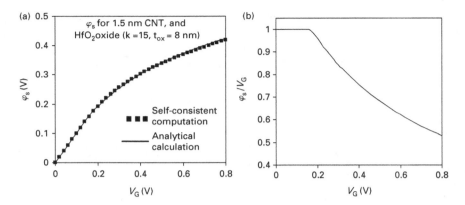

Fig. 8.3   Characteristic profile of the CNT surface potential. Analytical φ_s using best-fit parameters ($s = 1.41\ \mathrm{V}^{-1}$, $t = 0.76$) compared with self-consistent computation showing good agreement (peak error < 5 %). $C_{ox} \sim 264.96\ \mathrm{aF}\,\mu\mathrm{m}^{-1}$. (b) Analytical φ_s/V_G revealing that, at low voltages, V_G has maximum control of the surface potential which gradually weakens at higher voltages.

and t parameters used in Figure 8.3 ($s = 1.41\ \mathrm{V}^{-1}$, $t = 0.76$) are the best-fit values obtained empirically by curve fitting Eq. (8.13) to the numerical results. Using $m = 0.275/e$ and Eq. (8.12), the estimated analytical values for s and t are 1.52 V^{-1} and 0.83 respectively, which are within 10 % of the best-fit values.

In summary, the Taylor series approximation has revealed that the appropriate analytical form for the surface potential is Eq. (8.13) to first order, which shows very strong agreement with numerically computed values of φ_s. Additionally, the analytical parameters (s and t) can be determined from the Taylor series coefficients with increasing accuracy as C_q/C_{ox} becomes smaller than unity.

8.4   Ballistic theory of ohmic-contact CNFETs

Employing the analytical surface potential discussed in the previous section, we are now in a position to develop an analytical theory for CNFETs and thereby come to an understanding of their basic transistor properties and performance metrics. Of the four CNFETs identified in Table 8.1, the ballistic flavor with ohmic or transparent contacts is the most attractive because it allows for the smoothest flow of current, leading to superior performance compared with the other three flavors of CNFETs. As a result, our focus here will be on the ballistic CNFET with ohmic contacts. An added benefit is that the ballistic CNFET is actually the simplest CNFET to study mathematically,[12] leading to deep insights that can be extended

---

[12] While it is true that the ballistic CNFET is the easiest to study, it will not be fair if we failed to mention that it is by far the most challenging of CNFETs to fabricate for reasons related to the

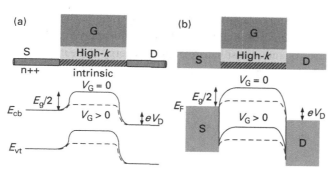

Fig. 8.4 Top-gated n-type CNFET device geometries considered within the ballistic model. (a) A CNFET with doped CNT ohmic contacts and its band diagram. (b) A self-aligned CNFET with ohmic metallic contacts and its band diagram ($k = 15$, $t_{ox} = 8$ nm for the experimental and analytical CNFETs analyzed in this section). Note that while electrons experience an ohmic contact, holes, on the other hand, experience a Schottky barrier.

to the other types of CNFET in Table 8.1. Invariably, we expect a successful transistor theory to capture all the pertinent parameters, including temperature and bandgap. Additionally, the external voltages by which the charge and current are controlled must be explicitly reflected in the transistor theory in order for the theory to be of any use.

Ballistic transport in CNTs applies when the channel length is much smaller than the electron mean free path l_m. In high-quality semiconducting CNTs operating around room temperature and at low energies, l_m is essentially determined by electron scattering from acoustic phonons which results in a mean free path of the order of 0.5 μm. Therefore, high-quality CNFETs with channel lengths ≲50 nm are naturally suited for ballistic operation. Notably, the ballistic CNFET can be realized in any of the device geometries shown in Figure 8.4. provided the channel length and contacts satisfy the expected criteria. Figure 8.4 shows the simplified device geometry of two top-gated CNFETs and associated band diagrams. In essence, the gate voltage controls the energy barrier for current flow between the source and drain electrodes, while the source and drain voltages control the source and drain Fermi levels respectively.

The current–voltage ($I-V$) relationship for a ballistic CNFET can be constructed from Planck's quantum postulate. An alternative derivation based on the DOS and group velocity can be found in previous studies.[13,14] We will focus on

difficulty of (i) making transparent contacts to semiconducting nanotubes and (ii) fabricating small channel lengths.

[13] K. Natori, Y. Kimura and T. Shimizu, Characteristics of a carbon nanotube field-effect transistor analyzed as a ballistic nanowire field-effect transistor. *J. Appl. Phys.*, **97** (2005) 034306.

[14] D. Akinwande, J. Liang, S. Chang, Y. Nishi and H.-S. P. Wong, Analytical ballistic theory of carbon nanotube transistors: experimental validation, device physics, parameter extraction, and performance projections. *J. Appl. Phys.*, **104** (2008) 124514.

the n-type transistor (or n-CNFET for short) operation, where the contributions of holes are considered negligible. The final results apply equally to p-CNFETs as well. Planck's quantum postulate relating the energy of the ith state to its frequency can be explicitly rewritten in terms of the current from the particles occupying the ith state:

$$E_i = hf_i = \frac{h}{2e}\frac{2e}{\tau_i} = \frac{h}{2e}I_i, \tag{8.14}$$

$$I_i = \frac{2e}{h}E_i, \tag{8.15}$$

where f_i and τ_i are respectively the frequency and period of particles in the ith state. For a system containing many states, the total current is the sum of the energy of each state and the associated probability of occupation of the state:

$$I = \frac{2e}{h}\sum_i F(E_i, E_F)E_i, \tag{8.16}$$

where $F(E, E_F)$ is the Fermi–Dirac statistic describing the probability that a state with energy E is occupied by electrons and E_F is the equilibrium Fermi energy:

$$F(E, E_F) = \frac{1}{1 + e^{\frac{E-E_F}{k_B T}}}. \tag{8.17}$$

In the limit that the states are very dense (i.e. the spacing between the discrete states becomes negligible), the sum can be converted to a Riemann integral:

$$I = \frac{2e}{h}\int F(E, E_F)\,dE. \tag{8.18}$$

In CNTs, there are right-moving carriers (electrons with positive velocities) and left-moving carriers (electrons with negative velocities); hence, the net current (commonly known as the drain current I_D) is the difference between the right- and left-moving carriers:

$$I_D = \frac{2e}{h}\int dE\, F(E, E_F) - \frac{2e}{h}\int dE\, F(E, E_F - eV_D). \tag{8.19}$$

The Fermi energy of the right-moving carriers is controlled by the source potential, while that of the left-moving carriers is controlled by the drain potential. We set the source potential to 0 V for reference and, as such, for a positive drain voltage, E_F at the drain electrode will be correspondingly pulled down (see Figure 8.4b for a visual illustration). Including the contributions of all the subbands in a semiconducting nanotube, the total drain current becomes the sum of the net current from each subband:

$$I_{D,tot} = \frac{2e}{h} \sum_j \int_{E_j - e\varphi_s}^{\infty} dE \, [F(E, E_F) - F(E, E_F - eV_D)], \qquad (8.20)$$

where j is the subband index and E_j is the energy minimum or bottom of the jth-subband (for reference, E_j for the first two subbands are shown in Figure 8.2). In general, Eq. (8.20) is computed by numerical integration for the transistor drain current. Fortunately, as a result of the analytical surface potential developed in the previous section, we can obtain a simplified closed-form expression for the drain current. The surface potential appropriate for a three-terminal device such as a transistor will be elucidated towards the end of this section. At low operating voltages characteristic of nanoscale technologies (supply voltages are typically ≤ 1 V), higher subbands can be ignored with negligible effect and Eq. (8.20), therefore, simplifies to the contribution of the first subband.[15]

$$I_D = \frac{4e}{h} \int_{E_{cb} - e\varphi_s}^{\infty} dE \, [F(E, E_F) - F(E, E_F - eV_D)]. \qquad (8.21)$$

The additional factor of 2 in the prefix is due to the first subband degeneracy. It follows that the integral yields an analytic formula for the drain current:

$$I_D = I_o \left\{ \ln \left[1 + e^{2e\varphi_s - E_g/2k_BT} \right] - \ln \left[1 + e^{2e\varphi_s - 2eV_D - E_g/2k_BT} \right] \right\}, \qquad (8.22)$$

where I_o is

$$I_o = \frac{4e}{h} k_B T = \frac{k_B T}{e R_q}, \qquad (8.23)$$

with R_q representing the usual quantum resistance ($R_q = h/4e^2 \sim 6.45$ kΩ). At 300 K, $I_o \sim 4$ μA. Equation (8.22) is the essential low-energy ballistic $I-V$ for a CNFET with transparent contacts as a function of the terminal voltages, material properties, and temperature.

Let us take a moment to contemplate on the functional dependencies in the formula and interpret its form. $k_B T$ comes about in the expression courtesy of the thermodynamic Fermi–Dirac distribution, R_q reflects the quantum transport of carriers, and E_g reflects the bandgap dependence. φ_s captures the band structure of the material (vis-à-vis the quantum capacitance) and also the effect of the external

---

[15] Furthermore, Guo and coworkers have shown by computation (for an intrinsic ballistic CNFET) that the first subband largely determines current under low-bias conditions. See J. Guo, A. Javey, H. Dai and M. Lundstrom, Performance analysis and design optimization of near ballistic carbon nanotube field-effect transistors. *IEEE IEDM Tech. Digest*, (2004) 703–6.

gate voltage in controlling the channel charge. V_D is the other external influence by which charge transport can be controlled. The first term is the current from the right-moving carriers, while the latter term is due to the left-moving carriers. An important observation is that the current from the backward carriers is significant in the linear region of the transistor $I_D - V_D$. The current from the forward carriers determines the saturation current, also known as the ON current I_{ON}, because at high V_D the latter term becomes negligible and the current saturates independent of V_D. We can obtain a rough estimate for the maximum I_{ON} by reasonably assuming that the maximum surface potential (in the low-energy ballistic regime) is slightly greater than the energy minima of the first subband, say $e\varphi_s \sim 1.5E_{cb} = 0.75E_g$. In this case, for d_t from 1–3 nm at room temperature, the maximum first subband I_{ON} is within a factor of 2 of 24 μA. For a p-CNFET, Eq. (8.22) is modified by reversing the polarity of φ_s and V_D.

$$I_D = I_o \left\{ \ln \left[1 + e^{-(2e\varphi_s + E_g)/2k_BT} \right] - \ln \left[1 + e^{(2eV_D - 2e\varphi_s - E_g)/2k_BT} \right] \right\}. \quad (8.24)$$

The surface potential appropriate for a three-terminal device such as the CNFET is slightly different from Eq. (8.1) defined in the previous section, which was based on a two-terminal device structure where the source and drain are shorted together. For the two-terminal device all the induced charges effectively travel in the same direction, leading to a displacement current, and the purpose of the surface potential is to account for all the induced charges in the nanotube. For the case of a three-terminal device such as the n-type CNFET, at equilibrium, half of the mobile electrons are right-moving carriers and the remaining half are left-moving carriers. Hence, the appropriate surface potential to be used for a CNFET (Eqs. (8.22) or (8.24)) must account for the charge carriers per direction in the nanotube. That is, for example, the surface potential in the first or latter term of Eq. (8.22) should account for only half of the induced charges, since those are the charges moving with positive or negative velocities respectively. In this case, the proper definition for the surface potential for CNFETs is

$$\varphi_s = \frac{C_{ox}}{C_{ox} + \frac{C_q(\varphi_s)}{2}} V_G = \frac{2C_{ox}}{2C_{ox} + C_q(\varphi_s)} V_G \quad (8.25)$$

where the factor of 2 represents the fact that only half the charges are traveling in any particular direction at any gate voltage with $V_D = 0$. This is, of course, similar to Eq. (8.1); as such, the analytical expression developed for the surface potential (Eq. (8.13)) can still be employed with C_{ox} replaced with $2C_{ox}$ in the expressions for the model parameters, s and t. Under excitation, the role of a finite V_D is to suppress the backward carriers in a CNFET.

An additional concern that needs to be incorporated in a general formulation for the surface potential is the flatband voltage V_{FB}.[16] All along we have been

---

[16] In a nutshell, the flatband voltage is the voltage applied to the gate to get a flat energy band diagram along the vertical cross-section of the device, and is given by $V_{FB} = \Phi_{MS}$ where Φ_{MS} is

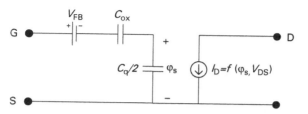

Fig. 8.5   An electrostatic intrinsic circuit model of the ballistic CNFET.

using V_G assuming flatband conditions at $V_G = 0$ V. In general, inclusion of the workfunction difference between the gate metal and the nanotube leads to

$$V_G = V'_G + V_{FB}, \tag{8.26}$$

where V'_G is the voltage under flatband conditions. The expression for V'_G should be used in the formula for the surface potential:

$$\varphi_s = (V_G - V_{FB}) \left[1 - \left(1 - \frac{1}{s|V_G - V_{FB}| + t} \right) H \left(1 - \frac{1}{s|V_G - V_{FB}| + t} \right) \right]. \tag{8.27}$$

For a non-zero voltage at the source electrode, V_{GS} and V_{DS} should be used instead of V_G and V_D respectively in the expressions for the drain current and surface potential. An intrinsic electrostatic (sometimes called large-signal) circuit model of the ballistic CNFET is shown in Figure 8.5.

The value of analytical theories depends on how successfully the theory corroborates experimental observations. For this purpose, the analytical ballistic theory, Eq. (8.24), is compared in Figure 8.6 with an experimental ballistic p-CNFET reported by Javey et al.[17] The model parameters used are identical to the values reported for the experimental CNFET ($d_t = 1.7$ nm, $E_g = 0.5$ eV, and device geometry shown in Figure 8.4b), including a parasitic source/drain metal resistance of 2 kΩ each, which is close to the reported estimate of ~1.7 kΩ. Parameter extraction of the flatband voltage from the experimental data yielded a value of $V_{FB} \sim 0.28$ V which is used in the comparison.[14] A modest agreement is observable in the active or current saturation region, while in the linear region the analytical ballistic theory significantly overestimates the current and conductance. The origin of the discrepancy can be identified upon closer examination of the band diagrams in Figure 8.4, and is elaborated upon in the next section.

the workfunction difference between the gate and the semiconductor. Basically, for a positive or negative V_{FB} the action of the gate is retarded or enhanced respectively in an n-CNFET, and vice versa for a p-CNFET. Enhancement (retardation) implies less (more) voltage is needed to turn ON the device.

[17] A. Javey, J. Guo, D. B. Farmer, Q. Wang, E. Yenilmez, R. G. Gordon, et al., Self-aligned ballistic molecular transistors and electrically parallel nanotube arrays. Nano Lett., **4** (2004) 1319–22.

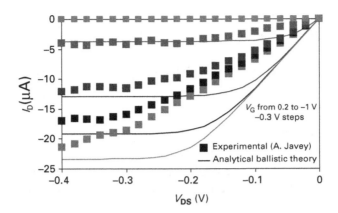

Fig. 8.6   Experimental results and analytical model predictions, Eq. (8.24), for a ballistic 50 nm p-type CNFET (identical to Figure 8.4b). $s = 1.6$ and $t = 0.82$ for the analytical ϕ_s. Agreement is good in the active region but weak in the linear region due to optical phonon scattering, which needs to be accounted for in the theory. (Experimental results are courtesy of Javey *et al.*[17])

8.5   Ballistic theory of CNFETs including drain optical phonon scattering

So far we have found the ballistic theory of CNFETs to be in modest agreement with experimental data in the current saturation or active region of the $I-V$ curve. However, a significant discrepancy exists in the linear region. A closer scrutiny of the band diagram of the ballistic CNFET proves useful in providing guidance as to the way forward to an improved theory. Figure 8.7 illustrates the conduction band diagram of an n-CNFET under two drain bias conditions. Figure 8.7a is at a low drain bias corresponding to the linear region, while Figure 8.7b corresponds to the saturation region. We can see visually that otherwise low-energy electrons traveling through the channel towards the drain will acquire significant energy at the drain end of the channel.[18] This forces us to re-examine our assumption of ballisticity. While it is true that electron transport through most of the channel is ballistic (a flat band signifies a ballistic channel without resistive voltage drop), at the drain end the prospect of optical phonon scattering characteristic of high-energy particles is certainly possible. In CNTs, the criteria to be satisfied for electron scattering by optical phonons is that (i) the length of travel through the well l_w, (see Figure 8.7) should be greater than the optical phonon mean free path ($l_{op} \sim 10$ nm) and (ii) the energy of the electrons must be greater than the critical optical phonon emission

---

[18] The kinetic energy of the electrons is the excess energy acquired beyond the conduction band energy minima.

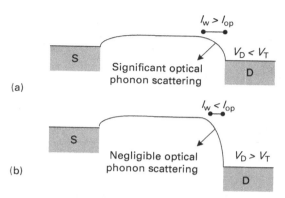

Fig. 8.7 Illustration of optical phonon scattering in a ballistic (n-type) CNFET. (a) At low V_D, the length of the triangle-like well l_w is comparable to the mean free path of optical phonon scattering ($l_{op} \sim 10$ nm), thereby reducing I_D. (b) At higher V_D, the well is sharpened, becoming gradually less than l_{op} and, hence, having a minimal effect on I_D. This device physics can be captured to first-order by the DOPS parameter α, which accounts for the reduced current.

energy ($E_{op} \sim 0.16$ eV; see Chapter 7). If these conditions are satisfied,[19] then optical phonon scattering will occur at the drain end of the channel and, hence, the current (and conductance) in the linear region will be reduced accordingly, corroborating experimental observations. At higher drain bias (saturation region), the triangular well becomes steeper,[20] with an ever-decreasing length, which implies that electron scattering by optical phonons becomes negligible as $l_w < l_{op}$, leading to minimal correction of the ballistic theory in the active region.

The scattering of high-energy electrons by optical phonons in an otherwise ballistic channel can be incorporated into the ballistic theory by a phenomenological parameter which is rightly called the drain optical phonon scattering (DOPS) parameter α. The justification of this parameter is as follows. Optical phonon scattering at the drain end essentially implies a localized resistance leading to an effective drain voltage across the ballistic portion of the channel that is less than the terminal drain voltage. In this light, we can construct a DOPS parameter to quantify the effective or reduced drain voltage. It follows that the effective drain voltage can be written phenomenologically as αV_D where $0 < \alpha \leq 1$, leading to a more realistic expression for the drain current:

$$I_D = I_o \left\{ \ln \left[1 + e^{(\pm 2e\varphi_s - E_g)/2k_B T} \right] - \ln \left[1 + e^{(\pm 2e\varphi_s \mp 2e\alpha V_D - E_g)/2k_B T} \right] \right\},$$
(8.28)

[19] It can be shown that the conditions can be satisfied in the linear region of the $I-V$ by employing semiconductor junction theory.

[20] In semiconductor device physics, the potential well is often considered a carrier depletion region and modeled as a capacitance. At higher drain voltages, the electrostatic attraction is greater, thereby pulling the electrons closer to the drain side, resulting in a steeper potential well.

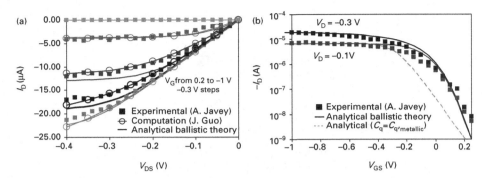

Fig. 8.8   (a) $I_D - V_{DS}$ including the DOPS parameter ($\alpha = 0.5$) showing strong agreement between analytical ballistic theory Eq. (8.28), experimental results, and quantum mechanical computation. (b) Likewise, $I_D - V_{GS}$ also results in strong agreement with experimental data in active, linear, and sub-threshold regions. The dashed line ($V_D = -0.1$ V) is the analytical prediction if $C_q = C_{q,\text{metallic}}$ is assumed, as is sometimes employed in compact modeling of semiconducting CNTs, leading to gross errors in the sub-threshold and active regions but which is fairly accurate in the linear region. (Experimental and quantum mechanical computation data are courtesy of A. Javey, J. Guo, et al.[17])

where the upper polarity is for n-CNFETs and the lower polarity is for p-CNFETs. In general, α should depend on both the gate and drain voltages, since they control the length and energy of the triangular well. For the greatest simplicity, let us consider the DOPS parameter a constant which is extractable from experimental data.[21] Returning to the experimental data of Javey et al.,[17] the maximum channel conductance value of $\sim 0.5/R_q$ was reported corresponding to $\alpha \sim 0.5$. Employing this value of α, we observe that the DOPS-corrected ballistic current, Eq. (8.28), is justified by the remarkably strong agreement with the experimental data in all major regions of transistor operation (sub-threshold, linear, and saturation), as shown in Figure 8.8. Detailed self-consistent quantum mechanical computations reported by J. Guo for the experimental device are also included for reference,[17] demonstrating the strength of the analytical ballistic theory in accurately reproducing the experimental and quantum mechanical results. The slight disagreement in the OFF state ($V_G \sim 0.2$ V in Figure 8.8b) is due to ambipolar electron current, which can be handled within the Schottky-barrier CNFET model developed in Section 8.8.

In order to provide clarity in the use of the term *ballistic transport* in transistors, it is necessary to make two distinctions:

(i) *Fully ballistic transport*: This is defined as transport with no acoustic or optical phonon scattering ($\alpha = 1$).

---

[21] The DOPS parameter α can be extracted from measurements of the drain to source channel conductance (g_{ds}), in the limit of $V_D \to 0$, and $V_G \to V_{DD}$ where V_{DD} is the drain supply voltage. In this limit, the conductance is $g_{ds} = \alpha/R_q$.

(ii) *Partially ballistic transport*: This is defined as transport with no acoustic phonon scattering but possibly some optical phonon scattering ($\alpha < 1$).

The DOPS effect is expected to apply to ballistic FETs in general, including silicon FETs.[22] Further research is warranted to investigate how the DOPS parameter scales with the device dimensions and semiconductor-contact dynamics, with the potential that the results of such investigations might lead to optimizing the device to minimize this effect. In the next section we will discuss and enumerate the performance parameters of the ballistic CNFET.

8.6   Ballistic CNFET performance parameters

Many of the basic parameters that characterize transistor performance can now be derived from the experimentally validated ballistic theory. This includes fundamental parameters such as the threshold voltage V_T, the transconductance ($g_m = \partial I_D / \partial V_G$), and the channel conductance ($g_{ds} = \partial I_D / \partial V_D$). In addition, the sub-threshold current I_{sth} and associated slope and the saturation or ON current I_{ON} are of routine interest.

The threshold voltage for an FET is defined as the gate voltage at which a sufficient amount of carriers are induced to create a conducting channel. Strictly speaking, in silicon FETs, V_T corresponds to the gate voltage where an inversion layer is induced at the interface between the gate oxide and the silicon substrate.[23] For an intrinsic semiconductor, the threshold voltage takes on a somewhat vague meaning because of the lack of carrier inversion, as is the case in conventional doped semiconductors. To circumvent this ambiguity, we can define a threshold voltage that does not rely on carrier inversion yet still have virtually the same properties characteristic of V_T in conventional FETs. Mathematically, we define V_T to be the gate voltage at which the gradient of the transconductance is a maximum, as shown in Figure 8.9a.[24] This definition is universally applicable to all CNFETs and in general to all FETs based on an intrinsic semiconductor channel. Moreover, the defined V_T can be directly determined from experimental measurements, avoiding altogether the extraction ambiguities related to small-dimensional effects in the conventional definition of V_T.

---

[22] K. Natori, Ballistic MOSFET reproduces current-voltage characteristics of an experimental device. *IEEE Electron Device Lett.*, **23** (2002) 655–7.

[23] For example, the channel of a silicon n-FET is normally doped with holes; therefore, a non-conducting path exists between the electron-doped S/D regions and the channel. To turn ON the device, a positive gate voltage exceeding V_T has to be applied to invert the channel from p-type to n-type, leading to the creation of a conducting path.

[24] This definition has been previously introduced for thin-oxide FETs. See H.-S. Wong, M H. White, T. J. Krutsick and R. V. Booth, Modeling of transconductance degradation and extraction of threshold voltage in thin oxide MOSFETs. *Solid-State Electron.*, **30** (1987) 953–68.

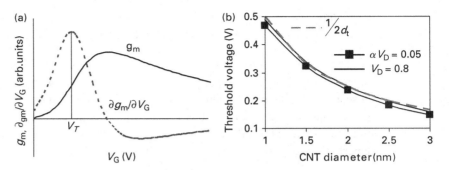

Fig. 8.9   (a) The graphical illustration of the definition of V_T for CNFETs with an intrinsic channel.
(b) V_T exactly computed from the analytical ballistic theory model for a ballistic 50 nm
n-type CNFET (derived from Figure 8.4) showing weak dependence on drain voltage and
a simple $\sim 0.5/d_t$ relation.

For an n-CNFET, the threshold voltage is approximately given by[14]

$$V_T \approx k_B T \ln \left\{ \frac{1}{12} e^{(E_g + e\alpha V_D)/2k_B T} \left[\sqrt{24 + e^{(e\alpha V_D)/k_B T}} - e^{(e\alpha V_D)/2k_B T} \right] \right\},$$
(8.29)

which can be further simplified (via a first-order Taylor series expansion) to reveal
a relationship that is directly proportional to the bandgap and, hence, inversely
proportional to the nanotube diameter (since $E_g \sim 1/d_t$):

$$eV_T \approx \frac{1}{2} E_g + \text{offset}.$$
(8.30)

This relationship is confirmed in Figure 8.9b (where the offset is simply set to zero),
showing the V_T computed exactly from the ballistic model compared with a simple
$V_T \sim 0.5/d_t$ (V nm) relation with good agreement. A suite of analytical formulas
for basic transistor performance parameters are shown in Table 8.2, which the
reader is strongly encouraged to derive for intellectual satisfaction. The parameters
are easily derivable using elementary algebra and calculus. The sub-threshold
current is the current when $V_G < V_T$ and is an increasingly important parameter
relevant to power dissipation and gate control in nanoscale electronics. Techniques
to extract the DOPS parameter and the bandgap or the flatband voltage are also
enumerated in the table.

Much of the excitement about CNFETs concerns their potential performance
advantages over silicon transistors for nanoscale digital, analog, and radio-
frequency (RF) applications. For this reason, several workers have benchmarked
the performance of relatively ideal CNFETs to silicon FETs, and the consensus outlook is that, qualitatively speaking, ballistic CNFETs offer superior performance to

Table 8.2. Expressions for the key metrics of CNFETs and some parameter extraction techniques

#	Analytical equations for ballistic n-type CNFET	Comments/conditions
(1)	$I_{ON} \approx \dfrac{k_B T}{e R_q} \ln \left[1 + e^{(2e\varphi_s - E_g)/2k_B T} \right]$	Saturation current in active region ($V_G > V_T$, $V_D > V_G - V_T$)
(2)	$I_{sth} \approx \dfrac{k_B T}{e R_q} \left[e^{(2eV_G - E_g)/2k_B T} - e^{(2eV_G - 2\alpha eV_D - E_g)/2k_B T} \right]$	Sub-threshold current with an ideal slope of 60 mV/decade at 300 K ($V_G < V_T$)
(3)	$V_G = V_T$, when $\dfrac{\partial g_m}{\partial V_G} = \text{maximum}$	Definition for V_T for intrinsic CNFETs
(4)	$V_T \approx k_B T \ln \left\{ \dfrac{1}{12} e^{(E_g + \alpha eV_D)/2k_B T} \left[\sqrt{e^{(\alpha eV_D)/k_B T} + 24} - e^{(\alpha eV_D)/2k_B T} \right] \right\}$	Analytical V_T in the limit of $\varphi_s \sim V_G$
(5)	$g_m = \dfrac{\partial I_D}{\partial V_G} = \left[\dfrac{1}{1 + e^{(2e\varphi_s - 2\alpha eV_D - E_g)/2k_B T}} - \dfrac{1}{1 + e^{(2e\varphi_s - E_g)/2k_B T}} \right] \dfrac{1}{R_q} \dfrac{\partial \varphi_s}{\partial V_G}$	Transconductance
(6)	$g_{ds} = \dfrac{\partial I_D}{\partial V_D} = \dfrac{\alpha}{R_q} \left[1 + e^{(E_g + 2\alpha eV_D - 2e\varphi_s)/2k_B T} \right]^{-1}$	Channel conductance
(7)	$g_{ds0} = \dfrac{\alpha}{R_q}$	Technique to extract α from g_{ds}($V_D \to 0$, $V_G \to V_{DD}$)
(8)	$2k_B T \ln \left(\dfrac{e R_q}{k_B T} I_{sth} \right) = 2e(V_G - V_{FB}) - E_g$	Technique to extract E_g or V_{FB} from measured I_{sth} ($V_D \to V_{DD}$)

silicon FETs for comparable channel lengths and electrical conditions.[14,15,25] The actual quantitative performance advantage depends on the modeling details and associated approximations. However, contemporary CNFETs that are frequently fabricated do not represent the ideal ballistic CNFET; and, furthermore, any performance benefits begin to diminish when parasitic capacitances are included in the device geometry.[26] At this moment, significant advances are required to routinely fabricate near-ideal nanotube FETs to realize their intrinsic electrical performance.

[25] J. Guo, S. Hasan, A. Javey, G. Bosman and M. Lundstrom, Assessment of high-frequency performance potential of carbon nanotube transistors. *IEEE Trans. Nanotechnol.*, **4** (2005) 715–21; L. Wei, D. J. Frank, L. Chang and H.-S. P. Wong, A non-iterative compact model for carbon nanotube FETs incorporating source exhaustion effects. *IEEE IEDM Tech. Digest*, (2009) 917–20.

[26] N. Patil, J. Deng, S. Mitra and H.-S. P. Wong, Circuit-level performance benchmarking and scalability analysis of carbon nanotube transistor circuits. *IEEE Trans. Nanotechnol.*, **8** (2009) 37–45.

8.7   Quantum CNFETs

Carbon nanotube FETs can exhibit very novel properties when working in the quantum capacitance limit, particularly in the active region of transistor operation. These novel properties, as we shall see shortly, are quite unique and not observable in conventional semiconductor FETs that possess a large DOS (such as silicon). For brevity, we refer to a CNFET that satisfies the quantum capacitance limit in all its regions of operation (sub-threshold, linear, and active) as a quantum FETs in order to distinguish its unique properties from the conventional or classical FET. The quantum capacitance limit entails that $C_{ox} \gg C_q$, and hence $\varphi_s \approx V_G$. Accordingly, the gate has maximum control of the electrical properties of the channel in a quantum CNFET.

To actually realize a quantum CNFET requires that the oxide capacitance be much greater, say about an order of magnitude greater, than the quantum capacitance. It is worthwhile obtaining an estimate of the oxide capacitance to satisfy this requirement. Continuing with our focus on low energies (i.e. only the first subband is considered), the maximum C_q of semiconducting nanotubes is about a factor of 2 greater than the quantum capacitance of metallic nanotubes. That is, maximum $C_q \approx 0.6 \text{ fF} \, \mu\text{m}^{-1}$ (see Chapter 6); therefore, C_{ox} needs to be $>6 \text{ fF} \, \mu\text{m}^{-1}$.[27] Substituting $\varphi_s = V_G$ into Eq. (8.28), the $I-V$ expression of a quantum CNFET is

$$I_D = I_0 \left\{ \ln \left[1 + e^{(\pm 2eV_G - E_g)/2k_B T} \right] - \ln \left[1 + e^{(\pm 2eV_G \mp 2e\alpha V_D - E_g)/2k_B T} \right] \right\}. \tag{8.31}$$

Consider the saturation current (say for an n-CNFET) given by

$$I_{ON} = I_0 \ln \left[1 + e^{(2eV_G - E_g)/2k_B T} \right] \approx \frac{V_G}{R_q} - \frac{E_g}{2eR_q}, \tag{8.32}$$

where the approximation is valid for typical gate voltages and bandgaps. For example, a CNT with a bandgap ~ 0.6 eV has a threshold voltage of ~ 0.3 V under flatband conditions. At room temperature, for $V_G > 0.35$ V, the exponential term is $\sim 54.6 \gg 1$. Since the current is linear with voltage, the transconductance is a constant:

$$g_m \approx \frac{1}{R_q} \tag{8.33}$$

---

[27] In principle, this can be realized by a 10 nm thin dielectric with $k > 350$. However, at present, it is not at all certain that this value of oxide capacitance on a CNT is achievable. Most higher k metal oxides, such as various types of titanates, have a thickness-dependent k which can exceed 1000 for thick films, but often drops below 200 for $t_{ox} < 50$ nm. At the same time, such higher k dielectrics have the undesirable effect of preferentially screening the drain field through the dielectric. See D. Frank et al. in footnote 35.

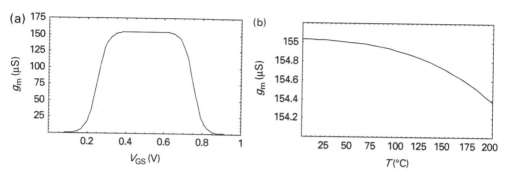

Fig. 8.10   Novel and highly desirable properties of the quantum CNFET. (a) The transconductance dependence on V_G revealing a flat response in the active region over a wide range of voltages (at 300 K). (b) Temperature profile of g_m, showing less than 1 % variation up to 150 °C. For this example, $E_g = 0.5$ eV, $V_D = 1$ V, $V_G = V_D/2$, and $\alpha = 0.5$.

This is a remarkable result in contrast with conventional FETs, where the transconductance in the active region can generally be expressed as a power series function of the small-signal gate voltage in the form $g_m \approx a_o + a_1 v_g + a_2 v_g^2 + \cdots$, where v_g is the small-signal gate voltage. The constant transconductance of a quantum CNFET is actually preferable, particularly for analog and RF circuits, where much effort is often devoted to suppressing the gate dependence of the transconductance in classical FETs that inevitably leads to signal distortion at large signal amplitudes.[28] In addition, the transconductance of the quantum CNFET also contains no explicit dependence on temperature, a property which is highly desirable in analog/RF circuits. The actual state of affairs regarding the transconductance profile of a quantum CNFET is shown in Figure 8.10. For this example, we have employed the actual $I-V$ expression in Eq. (8.31) to get an accurate outlook on the profile of g_m. The roll-off of the transconductance at higher gate voltages is because the semiconducting CNT is becoming more metallic-like.

8.8   Schottky-barrier ballistic CNFETs

Carbon nanotube FETs with Schottky (source/drain) metal contacts are frequently the type of CNFETs commonly encountered in practice not by intention or design, but simply because it is far easier to obtain Schottky contacts to semiconducting nanotubes instead of the more desirable ohmic kind. This is because it takes a great deal more diligent effort to engineer the metal–nanotube interface to be

---

[28] Common circuit techniques include ratiometric circuits and feedback or feedforward circuits. For an insightful discussion of distortion and linearity in analog/RF integrated circuits, see T. H. Lee, *The Design of CMOS Radio-Frequency Integrated Circuits*, 2nd edn (Cambridge University Press, 2004) Chapter 12.

transparent to charge carriers. Moreover, routine fabrication process conditions appear to conspire to favor Schottky barriers more often than not.[29] As a result, it is important to understand the device physics and $I - V$ characteristics of Schottky-barrier CNFETs. A welcome bonus from this endeavor is that the understanding gained deeply broadens our intuition about Schottky-barrier nanoscale transistors in general, including CNR transistors and semiconductor nanowire transistors. And in fact, understanding of Schottky-barrier device physics is necessary to understand (undesired) leakage current and the implications for ON/OFF current ratio in (metal) ohmic-contact ballistic CNFETs and similar 1D ballistic FETs, as will become evident in the course of the narrative.

The presence of Schottky barriers introduces substantial complexity in describing charge transport in the nanotube transistor owing to the inherently rich device physics. As such, it is quite easy to get lost or lose focus of the central device physics. This is especially true for the beginning student reader in this subject matter. For this reason, the discussion will be more of a gentle and simple introduction to Schottky-barrier nanotube transistors, with primary coverage of the basic transistor physics and the current–voltage characteristics. References will be provided along the way for advanced studies.

We begin our studies of Schottky-barrier CNFETs by exploring the bias-dependent band diagrams in order to obtain basic qualitative insight about charge transport. The use of band diagrams is essential in understanding and discussing charge dynamics in these transistors. For convenience, we focus our attention on an n-type transistor operation in the ballistic regime with a mid-gap Schottky barrier and zero flatband voltage,[30] conditions which are much easier to handle analytically and still provide us with the wealth of basic insights that are usually valid and can be adapted to explain charge transport under general conditions.

Figure 8.11 illustrates the band diagrams under different bias conditions, where the primary role of the gate voltage is to modulate the barrier thickness for electron tunneling, and the role of the (positive) drain voltage is to rigidly lower the barrier height at the drain electrode. The basic ideas regarding ballistic transport in ohmic-contact CNFETs apply here with the additional effects of quantum mechanical tunneling that must be properly accounted for. In a sense, the transistor can be considered as two Schottky barriers connected by a field-free region in the channel of the nanotube. We consider low-energy transport, which includes only the first subband.

Different bias conditions applied to the transistor can produce a diversity of charge transport phenomena, including source and drain electron tunneling in the

---

[29] Indeed, a search of published experimental nanotube transistors in one's favorite journals will unambiguously show that the vast majority (but not all) of CNFETs fabricated under different permutations of metal contacts, process conditions, and nanotube diameters produce Schottky-barrier CNFETs. Intuitively, metal workfunction and process and ambient conditions all play a role, but the degree of their influence in experimental devices remains unclear.

[30] In optimized device designs, the gate metal (and associated workfunction) is carefully selected to set the flatband voltage and, hence, the threshold voltage. The metal contacts are also chosen diligently to control the Schottky-barrier height.

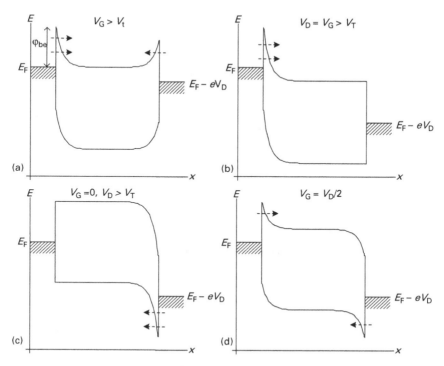

Fig. 8.11   Band diagrams of a ballistic n-CNFET with a mid-gap Schottky barrier under selected bias conditions (φ_{be} is the Schottky barrier height for electrons). (a) Band diagram in the linear region. The net current is the difference between electrons tunneling from the source and those tunneling from the drain. (b) The transistor ON state, where current is supplied from electrons tunneling from the source. (c) Band diagram in the OFF state. Note the valence band symmetry with the conduction band of (b). As a result, the current from holes tunneling from the drain will be identical to electron current in (b) for the same V_D. (d) Interesting bias conditions leading to symmetrical conduction and valence band profile and, hence, equal electron and hole currents.

linear region (Figure 8.11a) and electron source tunneling in the active or ON region (Figure 8.11b). Intriguingly, we discover it is difficult to turn off the transistor (OFF bias is conventionally defined as $V_G = 0$ V, and $V_D = V_{DD}$). In fact, for the device conditions under examination ($V_{FB} = 0$ V, and a mid-gap Schottky barrier), at $V_G = 0$ V, hole tunneling from the drain to the source results in current that is equal to the ON current when $V_G = V_D$. This instantly implies that the Schottky-barrier CNFET is fundamentally an ambipolar transistor, with electrons providing the bulk of the current above a critical gate voltage and holes responsible for the current below the critical V_G.[31] On deeper thought, we expect the

---

[31] These conclusions continue to hold for arbitrary Schottky-barrier heights, even including the special case of a vanishing φ_{bn} (a CNFET with transparent contacts which was the subject of previous sections). The effect of arbitrary Schottky-barrier heights is primarily to modify the magnitude of the ON and OFF currents and shift the position of the critical V_G. The overall

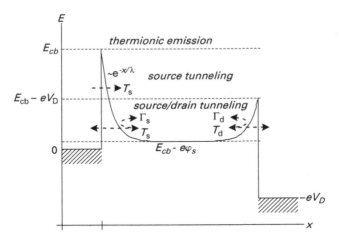

Fig. 8.12   The conduction band profile of a Schottky-barrier CNFET showing the source/drain
tunneling and reflection processes and the thermionic component of current from
high-energy electrons. $E_F = 0$ eV for reference.

critical gate voltage to correspond to the case when the electron and hole currents
must be equal. Indeed, we can construct a band diagram that is consistent with this
electron–hole current symmetry, as shown in Figure 8.11d, where the critical gate
voltage or symmetry point is determined to be $V_G = V_D/2$. With these qualitative
insights it is possible to forecast the general $I_D - V_G$ profile of the Schottky-barrier
CNFET without actually writing a single equation. In the OFF state, hole tunneling produces substantial current, which reduces as V_G increases. At the symmetry
point, electron tunneling equals hole tunneling and the current in the transistor
is at a finite minimum. For further increases in V_G, electron tunneling provides
increasing current, which reaches a maximum at $V_G = V_D$. Stated differently, for
a fixed V_D, the hole current increases with $V_D - V_G$ in exactly the same manner as
the electron current increases with V_G. All in all, a sketch of the $I_D - V_G$ is expected
to more or less reveal a V-shaped profile.

A quantitative semi-analytical model of transport in Schottky-barrier CNFETs
that is especially beneficial for nurturing intuition can be developed by employing the Landauer–Büttiker formalism where the Schottky barriers are treated as
scattering sites near the contacts.[32] Inside the channel, transport is ballistic and

ambipolar nature is still preserved. See J. Guo, S. Datta and M. Lundstrom, A numerical study of
scaling issues for Schottky-barrier carbon nanotube transistors. *IEEE Trans. Electron Devices*, **51**
(2004) 172–7. Note that, in CNFETs with a finite V_{FB}, the ambipolar feature may not be visible
within a limited gate voltage sweep, but inevitably becomes observable for a larger range.

[32] The theory of Schottky-barrier CNFETs often involves a set of assumptions and approximations.
In the present treatment, our goal is to create a simple physics-based model that predicts the
correct qualitative features while retaining reasonable accuracy.

equilibrium conditions apply. The contact themselves are considered to be reservoirs of electrons at (or near) thermal equilibrium and, hence, the use of the equilibrium Fermi–Dirac distribution is valid. Furthermore, we consider only the case where the mobile electrons have arbitrary or random phase. For this reason, constructive or destructive interference from multiple coherent scattering at the two tunneling barriers, which can lead to resonant tunneling phenomena,[33] can be neglected in the CNFET model. The neglect of interference is reasonable for currently achievable short channel lengths (tens of nanometers). However, at much shorter channel lengths roughly comparable to the thickness of the tunnel barriers, resonant tunneling effects become increasingly important.

We develop the Schottky-barrier CNFET model in simple stages of increasing detail, in line with the gentle introduction which is our driving philosophy. First, we state the formula for the Schottky-barrier ballistic current, then elucidate on the effective tunneling probabilities of mobile charges, and finally we will focus on developing an analytical expression for the tunneling probabilities applicable to CNFETs. This modeling strategy offers a seamless transition between Schottky-barrier CNFETs and ohmic-contact CNFETs discussed in the previous sections. The electron current I_e can be expressed following the Landauer–Büttiker formalism,[34]

$$I_e = \frac{4e}{h} \int_{E_{cb}-e\varphi_s}^{\infty} [T_{sd}(E)F(E, E_F) - T_{ds}(E)F(E, E_F - eV_D)]\, dE \qquad (8.34)$$

where T_{sd} is the source to drain transmission probability or coefficient of an electron-wave propagating from the source to the drain, successfully through both Schottky barriers, and vice versa for T_{ds}. This expression essentially has the same form has the previously discussed ballistic current integral, Eq. (8.21), but modified by the transmission probabilities. The transistor drain current is the sum of the electron and hole currents. We will explore the electron current for now in order to develop expressions for the tunneling probabilities; subsequently, it is a straightforward matter to write the expression for the hole current.

Let us consider an electron already in the channel, propagating to the drain. The pressing question is: What is the transmission probability of this electron propagating to the drain reservoir? Let us call this channel to drain transmission probability T_{cd}. Furthermore, electrons that do not tunnel through the barrier are reflected backwards. We can label the transmission and reflection coefficients at the source and drain as T_s, Γ_s and T_d, Γ_d respectively (see Figure 8.12). It follows

[33] J. Singh, *Quantum Mechanics: Fundamentals & Applications to Technology* (John Wiley, 1997).
[34] S. Datta, *Electronic Transport in Mesoscopic Systems* (Cambridge University Press, 1995).

that T_{cd} is a sequential sum of a series of scattering events:

$$T_{cd} = T_d + \Gamma_d\Gamma_s T_d + (\Gamma_d\Gamma_s)^2 T_d + (\Gamma_d\Gamma_s)^3 T_d + \cdots + (\Gamma_d\Gamma_s)^n T_d,$$

$$T_{cd} = T_d \left[1 + \sum_i^n (\Gamma_d\Gamma_s)^i \right]. \tag{8.35}$$

Likewise, the total probability that the electron will be reflected into the source reservoir Γ_{cs} is a sum of events:

$$\Gamma_{cs} = \Gamma_d T_s + \Gamma_d^2\Gamma_s T_s + \Gamma_d^3\Gamma_s^2 T_s + \cdots + \Gamma_d^n\Gamma_s^{n-1} T_s,$$

$$\Gamma_{cs} = \frac{T_s}{\Gamma_s} \sum_i^n (\Gamma_d\Gamma_s)^i. \tag{8.36}$$

Owing to charge conservation (i.e. the electron can be found either at the source or drain with a certain probability), the transmission and reflection coefficients are related by

$$T_{cd} + \Gamma_{cs} = 1. \tag{8.37}$$

Substituting Eq. (8.36) for Γ_{cs}, the channel to drain transmission probability in terms of T_s and T_d simplifies to

$$T_{cd} = 1 - \frac{T_s}{\Gamma_s} \sum_i^n (\Gamma_d\Gamma_s)^i = 1 - \frac{T_s}{\Gamma_s} \frac{(T_{cd} - T_d)}{T_d},$$

$$T_{cd} = \frac{T_d}{T_s + T_d - T_s T_d}. \tag{8.38}$$

Therefore, it follows that the source to drain transmission probability is T_{cd} multiplied by the source transmission probability:

$$T_{sd} = \frac{T_s T_d}{T_s + T_d - T_s T_d} = T_{eff}. \tag{8.39}$$

Note that the symmetry involving T_s and T_d in the expression immediately implies that $T_{ds} = T_{sd}$. To retain this symmetry in the symbolic notation, T_{eff} will be used hereafter. The electron current can now be stated as

$$I_e = \frac{4e}{h} \int_{E_{cb}-e\varphi_s}^{\infty} T_{eff}(F_s - F_d)\,dE, \tag{8.40}$$

where the shorthand F_s and F_d are used to represent the Fermi–Dirac distribution at the source and drain respectively. Similarly, the hole current is given by

$$I_h = \frac{4e}{h} \int_{-E_{cb}-e\varphi_s}^{-\infty} T_{eff}(F_d - F_s)\,dE. \tag{8.41}$$

The total drain current is $I_D = I_e + I_h$.

We can now go one step further and expand I_e for the usual operating biases of an n-CNFET ($V_G \geq 0$, $V_D \geq 0$) regardless of the specific form of T_{eff} in order to demonstrate how the limits of the integral need to be handled. Since the Schottky barrier only exists for a finite range of electron energies, the restrictions on the allowed values of the source transmission probabilities are (see Figure 8.12 for visual illustration)

$E \leq E_{cb} - e\varphi_s,$ $T_s = 0$ (electrons below the bottom of the band have

no states to tunnel into);

$E_{cb} - e\varphi_s < E < E_{cb},$ $T_s = T_s(E)$ (the energy range where the

Schottky-barrier exists);

$E \geq E_{cb},$ $T_s = 1$ (thermionic emission of high-energy electrons

above the barrier).

Similarly, the drain transmission probabilities must satisfy the following restrictions when $E_{cb} - e\varphi_s < E_{cb} - eV_D$:

$$E \leq E_{cb} - e\varphi_s, \qquad\qquad T_d = 0;$$
$$E_{cb} - e\varphi_s < E < E_{cb} - eV_D, \quad T_d = T_d(E);$$
$$E \geq E_{cb} - eV_D, \qquad\qquad T_d = 1,$$

otherwise, when $E_{cb} - e\varphi_s \geq E_{cb} - eV_D$, $T_d = 1$, because the Schottky barrier no longer exists at the drain contact.

For $E_{cb} - e\varphi_s < E_{cb} - eV_D$, the electron current can be expanded into three integrals:

$$I_e = \frac{4e}{h}\left[\underbrace{\int_{E_{cb}-e\varphi_s}^{E_{cb}-eV_D} T_{eff}(F_s - F_d)\,dE}_{\text{both barriers present}} + \underbrace{\int_{E_{cb}-eV_D}^{E_{cb}} T_s(F_s - F_d)\,dE}_{\substack{\text{only source barrier} \\ \text{present}}} + \underbrace{\int_{E_{cb}}^{\infty} (F_s - F_d)\,dE}_{\substack{\text{thermionic} \\ \text{emission}}}\right], \tag{8.42}$$

while for $E_{cb} - e\varphi_s \geq E_{cb} - eV_D$, the electron current is

$$
I_e = \frac{4e}{h} \left[\int_{E_{cb}-e\varphi_s}^{E_{cb}} T_s(F_s - F_d) \, dE + \int_{E_{cb}}^{\infty} (F_s - F_d) \, dE \right]. \tag{8.43}
$$

The key point from this exercise is the need to handle the limits of the integral(s) thoughtfully to reflect the changing existence of the Schottky barriers.

In order to quantify the impact or effect of the Schottky barriers in reducing the current when the transistor is ON, we can introduce a dimensionless metric appropriately termed the Schottky-barrier impact (SBI) factor Λ defined as

$$
\Lambda(V_{G,D} = V_{DD}) = \frac{I_{D,\text{Ohmic}} - I_{D,\text{SB}}}{I_{D,\text{Ohmic}}} = \frac{\int_{E_{cb}-e\varphi_s}^{E_{cb}} (1 - T_s)(F_s - F_d) \, dE}{\int_{E_{cb}-e\varphi_s}^{\infty} (F_s - F_d) \, dE}, \tag{8.44}
$$

where $I_{D,\text{Ohmic}}$ is the ballistic current in a CNFET with transparent contacts and $I_{D,\text{SB}}$ is the corresponding current in the presence of Schottky barriers. Note that, in the ON state, the electron current given by Eq. (8.43) should be used for $I_{D,\text{SB}}$. For values of Λ approaching unity, the Schottky barrier has minimal impact on the ON current, while for $\Lambda \to 0$, the Schottky barrier substantially degrades the ON current. If Schottky barriers are inevitable in a particular FET, to the goal is to engineer the barriers so as to achieve Λ values close to unity.

The final part of the analysis is to model the tunneling of electrons through the Schottky barriers with the aim of obtaining an analytical expression for T_s and T_d. The position-dependent energy band potential profile for mid-gap Schottky barriers (see Figure 8.12) can be modeled with exponential functions to first order, described by

$$
E_c(x) = E_{cb} + e\varphi_s \left[e^{-x/\lambda} + e^{(x-L)/\lambda} - e^{-L/\lambda} - 1 \right]
$$
$$
+ eV_D \left[e^{-L/\lambda} - e^{(x-L)/\lambda} \right], \quad 0 \leq x \leq L, \tag{8.45}
$$
$$
E_v(x) = E_c(x) - E_g, \tag{8.46}
$$

where E_c and E_v are the position-dependent conduction band minima and valence band maxima respectively and L is the channel length. The reader is welcome to show that the expressions for the potential profiles are indeed valid given that the reference (zero) potential is the source E_F. λ is a length scale on which potential variations occur; in other words, a measure of the Schottky barrier thickness. The concept of λ is rooted in evanescent-mode analysis, which in essence decouples the vertical electrical field from the horizontal field and enables simple exponential expressions for the potential variation.[35] An accurate theory for λ has yet to be developed for CNFETs; however, some analysis has revealed that it is proportional and comparable to the oxide thickness.[36] In the meantime, it can be considered a phenomenological fitting parameter until further research produces an experimentally verifiable theory.

An analytical expression for the transmission probabilities is achievable within the Wentzel–Kramers–Brillouin (WKB) approximation by means of the potential profile expressions and the effective mass theorem, which considers the low-energy mobile electrons as free electrons with an effective mass m^* that accounts for their solid-state environment. The details of the WKB approximation are deferred to the many introductory quantum-mechanical textbooks that provide rigorous treatment.[33] The effective-mass approximation is likely the weakest assumption so far. Fortunately, published analysis suggests that the effective-mass approximation is reasonably accurate for tunneling in carbon-based transistors with Schottky barrier heights less than ~ 0.5 eV.[37] For mid-gap Schottky barriers this will include nanotubes with bandgaps up to ~ 1 eV (down to $d_t \sim 0.9$ nm). The primary benefit of the effective-mass approximation within the WKB formalism is the inherent simplicity and the accessible qualitative insights. With this in mind, the WKB formulas for the transmission probabilities are (subject to the energy restrictions stated previously)

$$\ln(T_s(E)) = -\frac{4\lambda\sqrt{2m^*}}{\hbar}\left[\sqrt{E_{cb}-E} - \sqrt{E-(E_{cb}-e\varphi_s)}\,\tan^{-1}\left(\frac{E_{cb}-E}{E-(E_{cb}-e\varphi_s)}\right)\right],$$

$$(8.47)$$

$$\ln(T_d(E)) = -\frac{4\lambda\sqrt{2m^*}}{\hbar}$$
$$\times\left[\sqrt{E_{cb}-eV_D-E} - \sqrt{E-(E_{cb}-e\varphi_s)}\,\tan^{-1}\left(\frac{E_{cb}-eV_D-E}{E-(E_{cb}-e\varphi_s)}\right)\right].$$

$$(8.48)$$

In general, the transmission probabilities have a quasi-exponential dependence on energy, with values approaching unity for tunneling electrons close to the top of the Schottky barrier. These formulas are valid for an n-CNFET with an equilibrium mid-gap Schottky barrier under normal operating bias conditions ($V_G, V_D \geq 0$). For p-CNFETs, and arbitrary Schottky barrier heights, the formulas can be adapted

[35] For in-depth discussions regarding λ, see D. J. Frank, Y. Taur and H.-S. P. Wong, Generalized scale length for two-dimensional effects in MOSFETs. *Electron Device Lett.*, **19** (1998) 385–7; and S. H. Oh, D. Monroe and J. M. Hergenrother, Analytic description of short-channel effects in fully-depleted double-gate and cylindrical, surrounding-gate MOSFETs. *IEEE Electron Device Lett.*, **21** (2000) 443–7.

[36] A. Hazeghi, T. Krishnamohan and H.-S. P. Wong, Schottky-barrier carbon nanotube field-effect transistor modeling. *IEEE Trans. Electron Devices*, **54** (2007) 439–45.

[37] P. Michetti and G. Iannaccone, Analytical model of 1D carbon-based Schottky-barrier transistors. *IEEE Trans. Electron Devices*, **57** (2010) 1616–25.

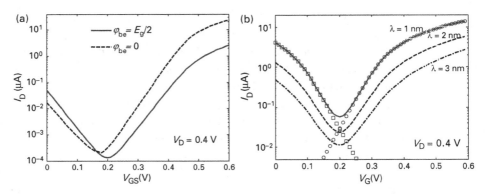

Fig. 8.13   n-CNFET characteristics. (a) I_D-V_G with transparent contacts ($\varphi_{be} = 0$) and mid-gap Schottky barriers ($\varphi_{be} = E_g/2$). Ambipolar characteristics exist in both cases. $d_t \sim 1$ nm ($E_g \sim 0.9$ eV), and $\lambda \sim 1.1$ nm, corresponding to a thin gate oxide (say <5 nm). (b) The consequences of increased Schottky-barrier thickness ($\sim\lambda$). Increasing λ reduces the ON current (approximately) exponentially and also degrades the sub-threshold slope. For this example, $\lambda = 1$ nm, 2 nm, and 3 nm results in sub-threshold slopes of $\sim$70 mV/decade, $\sim$82 mV/decade, and $\sim$100 mV/decade respectively. For $\lambda = 1$ nm, the open circles and squares are the electron and hole current contributions respectively. $d_t \sim 1.5$ nm.

in a straightforward manner. The band-minima effective mass was derived in Chapter 5 and is recalled here for convenience:

$$m^* = \frac{4\hbar^2}{3\gamma a^2} \frac{E_g}{2\gamma + E_g} \tag{8.49}$$

where the band structure constants γ and a are numerically $\sim$3.1 eV and $\sim$2.46 Å respectively.

Figure 8.13 is a graph of the characteristic $I-V$ profile of Schottky-barrier ballistic CNFETs revealing ambipolar transport. The minimum current occurs at $V_G = V_D/2$,[38] and corresponds to the band diagram of Figure 8.11d, as previously noted. Likewise, charge transport corresponding to $V_G = V_D$ and $V_G = 0$ V can also be explained visually by Figure 8.11b and Figure 8.11c respectively. The Schottky barriers usually reduce and limit the ON current compared with an identical CNFET with transparent contacts. Moreover, thicker barriers result in significant current reduction and degraded sub-threshold slope. Likewise, the transistor conductance also suffers from thicker barriers (i.e. thicker oxides) or taller barriers.[36] From an application point of view, the substantial OFF current is generally unacceptable; hence, in practice, gate workfunction engineering is

---

[38] This bias point is actually the optimum point for electron–hole recombination which produces maximum optical emission. See J. A. Misewich, R. Martel, Ph. Avouris, J. C. Tsang, S. Heinze and J. Tersoff, Electrically induced optical emission from a carbon nanotube FET. *Science*, **300** (2003) 783–6.

required to produce the necessary flatband voltage that shifts the $I-V$ curve to the left such that the minimum current is ideally at $V_G = 0$ V. For completeness, we note that decreasing the bandgap for a fixed V_D or increasing V_D (or the drain power supply) for a fixed bandgap results in appreciably greater increase in the minimum current than in the ON current; consequently, the ON/OFF current ratio is significantly compromised.[39] To sum up, the non-trivial challenges in obtaining suitable transistor characteristics over a moderate range of biases calls for great care in optimizing the CNFET for maximum performance, and also motivates the development of unipolar CNFET transistors, which are discussed in the next section.

As a final note, all along, ballistic transport has been assumed without explicit discussion of the appropriate mean free path. Does the mean free path due to acoustic phonons ($l_{ac} \sim 1$ μm) apply? Or is the much shorter optical phonon mean free path ($l_{op} \sim 10$ nm) the dominant length scale for Schottky-barrier CNFETs? A close inspection of Figure 8.11 or Figure 8.12 shows that a non-negligible number of the tunneling charge carriers are high-energy electrons, especially for those close to the top of the barrier, where the transmission probabilities are approaching unity. It is reasonable to expect that these electrons will have energies above the critical optical phonon energy $E_{op} \sim 0.16$ eV (as discussed in Chapter 7) and are susceptible to optical phonon scattering. Therefore, the effective mean free path will be a carrier-weighted average of the acoustic phonon mean free path (due to low-energy electrons) and the optical phonon mean free path (due to high-energy electrons). Because the average is always smaller than the longest of the mean free paths, we can conclude that the effective mean free path will be shorter than l_{ac} (compared with CNFETs with ohmic contacts, where l_{ac} applies). For this reason, the channel length of Schottky-barrier ballistic CNFETs should certainly be much less than l_{ac} (more so than for ballistic CNFETs with ohmic contacts), and might have to be closer to l_{op} in the ON state, where an increasing fraction of the current is due to high-energy electrons. Even then, much of the device physics elucidated is relevant if scattering effects are considered.

8.9   Unipolar CNFETs

There is great interest in minimizing or suppressing the drain-induced leakage current in the transistor OFF state noticeable in ambipolar-type CNFETs, including both the Schottky barrier and the metal ohmic-contact versions. The reason is quite obvious: substantial OFF currents leads to unacceptable (useless) power consumption, which is one of the primary concerns for nanoscale technology. One

---

[39] M. Radosavljević, S. Heinze, J. Tersoff and Ph. Avouris, Drain voltage scaling in carbon nanotube transistors. *Appl. Phys. Lett.*, **83** (2003) 2435–7.

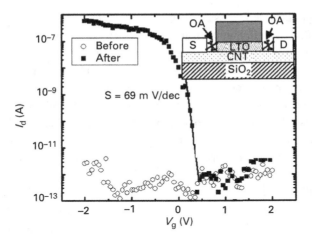

Fig. 8.14   Room-temperature transfer characteristics of a top-gated CNFET showing unipolar behavior after chemical doping of the nanotube contact extension regions (OA is the charge transfer doping agent). The inset shows a cross-section of the device. Reprinted with permission from J. Chen *et al.*[40]. Copyright (2005), American Institute of Physics.

technique to mitigate the OFF current is to engineer the gate flatband voltage such that the transistor minimum current symmetry point is at $V_G = 0$ V. However, this technique is largely insufficient, since the minimum current at the symmetry point has an exponential dependence on the drain voltage and is frequently appreciable enough to prevent adequate ON/OFF current ratios.[39] Therefore, generally robust techniques are desirable for either suppressing ambipolar transport or eliminating the OFF leakage current altogether. One promising technique to obtain unipolar-like CNFETs is to engineer the nanotube device to be similar to a conventional metal–oxide–semiconductor FET (MOSFET) with chemically doped contacts while retaining an intrinsic channel. Figure 8.14 highlights the transfer characteristics of an experimental CNFET with doped source/drain contact regions showing unipolar behavior over a wide range of gate voltages and good switching performance. In this device, the unipolar characteristic is attributed primarily to the Schottky barrier reduction courtesy of the workfunction modification of the source and drain electrodes under chemical doping.[40]

At the moment, there are several challenges facing the routine fabrication of chemically doped unipolar carbon transistors, including the synthesis of repeatable room-temperature air-stable p-type and n-type dopants for complementary transistor applications. In addition, the doping concentration must be precisely controlled and the dopants robust against any further device processing (such as annealing or surface passivation), and must prove reliable over repeated transistor

40 J. Chen, C. Klinke, A. Afzali and Ph. Avouris, Self-aligned carbon nanotube transistors with charge transfer doping. *Appl. Phys. Lett.*, **86** (2005) 123108.

operation in the long term. Research in this direction is ongoing for CNFETs on conventional and flexible substrates, and the desirable low-power advantages of unipolar-like carbon transistors suggests that this pursuit will remain active for the foreseeable future. In conclusion, it is gratifying that the intuition and understanding gained from the device physics of ambipolar CNFETs extend to chemically doped unipolar CNFETs with the pleasant benefit of simply neglecting ambipolar (leakage) current. That is, electron current can be neglected in p-type devices and vice versa.

8.10 Paradigm difference between conventional 2D MOSFETs and ballistic 1D FETs

Charge transport in 1D ballistic FETs with transparent contacts represents a different paradigm compared with conventional 2D MOSFETs. In this sense, ballistic CNFETs with transparent contacts embody an ideal 1D FET where its narrow cylindrical nature confines mobile electrons to travel either forward or backwards in one direction, while a conventional Si-MOSFET is an example model of a 2D FET, where mobile electrons are free to move in a 2D plane. Furthermore, at comparable length scales of practical relevance ($\sim$10–100 nm), CNFETs afford ballistic transport whereas diffusive transport dominates in a 2D MOSFET.[41] For these reasons, charge transport in 1D FETs such as CNFETs reflects a different paradigm even though the basic planar device geometry and current–voltage characteristics are very similar. We must emphasize that other examples of ballistic 1D FETs that experience this paradigm shift include ballistic nanowire and nanoribbon FETs of sufficiently small width and ideal edge passivation.

Let us consider an ideal (textbook-like) long-channel MOSFET and a ballistic CNFET to elucidate the similarities and differences in the main regimes of carrier transport, including sub-threshold, linear, and active regions. In the sub-threshold region, current in both devices is dominated by carriers above the gate-controlled energy barrier, whose increase in the ideal case is fundamentally determined by Fermi–Dirac thermodynamics. This is the origin of the room-temperature 60 mV/decade (i.e. 60 mV increase in V_G leads to a decade increase in I_D) sub-threshold slope in transistors that operate by diffusion of over-the-barrier electrons. In the linear region, charge transport in MOSFETs is by the process of drift, which increases with the horizontal electric field (set by the drain voltage V_D) and the carrier mobility u_{eff}. The carrier mobility is a useful parameter (related to the mean free path) that in essence quantifies the average scattering of carriers in the

---

[41] The longer mean free path of (low-energy) electrons in carbon nanotubes is arguably in part because of its 1D nature, which leads to reduced DOS for backscattering, though we have yet to see a rigorous exposition along this line.

Table 8.3. Paradigm difference between conventional 2D FETs and ballistic 1D FETs

MOSFET (2D)		CNFET (1D)	
$I_{ON} \approx u_{eff} \frac{C_{ox}W}{2L} (V_G - V_T)^2$		$I_{ON} \approx \frac{k_B T}{eR_q} \ln\left(1 + e^{\frac{2e\varphi_s - E_g}{2k_B T}}\right)$	
Parameter	Origins	Parameter	Origins
(i) u_{eff}	drift velocity	(i) φ_s	gate coupling
(ii) I/L	drift velocity	(ii) E_g	bandstructure
(iii) C_{ox}/W	charge sheet	(iii) R_q	quantum transport
(iv) V^2	drift velocity	(iv) $k_B T$	Fermi–Dirac
	charge sheet		thermodynamics

solid as they traverse the channel, and is a material-dependent property. For ballistic CNFETs, current in the linear region is determined by the difference between right-moving (from the source) and left-moving carriers (from the drain). In the active region, MOSFET current saturates as a result of the *pinch-off* of the channel and, hence, the drift velocity is no longer dependent on the drain voltage. For CNFETs, current saturation is due to *carrier density invariance* or *saturation* because of the vanishing left-moving carriers from the drain. In other words, only the right-moving carriers from the source exist in the channel; hence, the current no longer depends on the drain voltage.

Based on the paradigm difference, it is possible to extrapolate and offer some thoughts on comparative trends in channel length scaling by examining the origins of all the parameters in the ON state current expression for both devices as enumerated in Table 8.3. We note that the ON current in Si-MOSFET is largely controlled by the drift velocity, while that in a CNFET explicitly reveals many of the fundamental physical properties of electrons in the solid-state material, such as the bandgap, Fermi–Dirac thermodynamics, and quantum conductance. As such, electrical properties as a function of scaling are invariably different. For example, at high fields or for short channels, the drift velocity gradually saturates to a constant, which leads to a different ON current expression for MOSFETs that is not dependent on mobility or channel length and is quasi-linear with respect to gate voltage. Overall, the Si-MOSFET does not scale gracefully to short channels.[42] In contrast, the ON current for CNFETs with transparent contacts reveals no direct vulnerability to high-field effects. Indirect arguments can nonetheless

---

[42] This is the reason why there have been many great engineering efforts to make up for the degradation due to short-channel effects, including straining the channel to boost the current drive and introducing high-mobility materials to serve as the channel, such as high-mobility III–V semiconductors in place of the lower mobility silicon channel in an otherwise silicon-based VLSI technology.

be made for some short-channel effects in CNFETs. A case in point is the consequences of the increasing fraction of the channel length that is associated with the length scale of optical phonon scattering (i.e. DOPS) at the drain contact, as was highlighted previously. The short channel effects due to 2D electrostatic coupling are, of course, common to both devices owing to their similar device geometry. Additional non-idealities include gate leakage current due to thin oxides, possible *band-to-band* tunneling for low-bandgap materials, and direct source-to-drain tunneling for ultra-short channel lengths. All in all, CNFETs have better immunity to short channel effects and are expected to scale more gracefully to about 10 nm or so. Below 10 nm channel length, additional considerations, including the impact of the precise scaling of λ, and quantum mechanical effects, such as resonant tunneling phenomena, have to be examined carefully. Certainly, more systematic experimental studies would obviously be most welcome in providing a clearer picture of the scaling trend.

8.11 Summary

This chapter has provided an introduction to CNFETs with a primary emphasis on the understanding of the basic electronic properties for the purpose of developing intuition and insights. Some of the key device physics elucidated in this chapter include:

(i) Ballistic transport in devices with transparent contacts offers the maximum current.
(ii) Nanotubes operating in the quantum capacitance limit have unique properties, including voltage- and temperature-independent transconductance.
(iii) Ambipolar transport is characteristic of nanotube transistors that have direct metal contacts (either Schottky barrier or ohmic).
(iv) Schottky barriers in general degrade the sub-threshold slope, reduce the conductance, and limit the ON current. Sufficiently thin Schottky barriers have properties close to CNFETs with transparent contacts.
(v) Unipolar-like nanotube transistors are achievable with doped contact regions.
(vi) There exists a paradigm difference between charge transport in conventional MOSFETs and CNFETs.

Another insightful treatment of CNFETs (and nanoscale silicon FETs) can be found in the work of Lundstrom and Guo.[43] Much of the device physics of CNFETs readily extend to GNR transistors, and the analytical ballistic theory can be adapted

---

[43] M. Lundstrom and J. Guo, *Nanoscale Transistors: Device Physics, Modeling and Simulation* (Springer, 2006).

to GNR transistors by taking into account their bandgap and the number of degenerate subbands for low-energy transport. In general, the qualitative insights regarding ballistic CNFETs also apply to diffusive-type CNFETs. However, an accurate quantitative description has to properly take into account charge scattering and non-equilibrium effects, such as carriers that may have gained sufficient energy from the electric field or from absorbing a phonon and are no longer at thermal equilibrium with the reservoir contacts. Such quantitative theories are often founded on the quantum-mechanical non-equilibrium Green's function (NEGF) formalism, which is much more sophisticated but also computationally (time) expensive. The interested reader will find the textbook by S. Datta a useful introduction to NEGF formalism.[34] NEGF formalism is also very important in benchmarking the accurary of analytical and compact nanotube transistor models, as we showed with the analytical ballistic theory. An emerging nanotube transistor device that has not been discussed is the tunnel CNFET that employs band-to-band tunneling as the main voltage-controlled current mechanism.[44] This emerging device can offer a steeper sub-threshold slope, resulting in lower power dissipation. Transistors based on band-to-band tunneling are currently a matter of research.

    Finally, it is highly desirable to have CNFETs with multiple nanotubes operating in parallel so as to have control of the ON current and be able to supply arbitrary amounts, from micro-amperes to milli-amperes depending on the application requirements.[26] For this reason, the study and application of CNFETs motivates the ongoing research in growing aligned arrays of nanotubes. It is expected that detailed analysis of the performance of CNFETs with multiple nanotubes will continue to inform the research on synthesis in order to achieve acceptable nanotube densities (CNTs/micrometer) and narrow statistical distribution of diameters for high-performance applications.

8.12   Problem set

All the problems are intended to exercise and refine analytical techniques while providing important insights. If a particular problem is not clear, the reader is encouraged to re-study the appropriate sections. Invariably, some problems will involve making reasonable approximations or assumptions beyond what is already stated in the specific question in order to obtain a final answer. The reader should not consider this frustrating, because this is obviously how problems are solved in the real world.

8.1. Quantum capacitance versus electrostatic capacitance in the sub-threshold region.

[44] J. Appenzeller, Y.-M. Lin, J. Knoch and Ph. Avouris, Band-to-band tunneling in carbon nanotube field-effect transistors. *Phys. Rev. Lett.*, **93** (2004) 196805.

It was noted in the text that, at sufficiently low gate voltages, the gate coupling function is approximately unity. In this exercise, we wish to determine how reasonable such an approximation is. For gate oxides with a thickness ranging from 5 nm to 500 nm and dielectric constants anywhere from 4 to 20, compare the electrostatic capacitance with the quantum capacitance of a CNT with diameters within 1 nm to 3 nm, operating at very low gate bias. Is the approximation $\varphi_s \approx V_G$, valid for the conditions above? (*Hint*: Use of the expression for the non-degenerate quantum capacitance will spare the reader a lot of tedious algebra).

8.2. Diffusive CNFET with ohmic contacts.

Ballistic CNFETs with ohmic or transparent contacts have been discussed extensively throughout the text. However, in practice, most nanotube transistors do not enjoy this ideal condition. Largely due to fabrication challenges, the channel lengths of typical CNFETs are longer than the electron mean free path; hence, scattering is inevitable at room temperature. Such a CNFET is diffusive, in contrast to the ballistic ideal. The question then is: How can we adapt the analytical ballistic theory to account for scattering in a diffusive CNFET? This exercise is intended to develop a simple analytical diffusive theory that is useful for qualitative insights, by extending in a phenomenological manner the analytical ballistic theory.

Let us consider an n-CNFET where the channel is diffusive and is contacted by ideal transparent metals. Since the only difference between this diffusive CNFET and the ideal ballistic CNFET is the scattering present in the channel, our singular challenge is to find an analytical way to modify the expression for the first subband ballistic I_D that accounts for scattering.

(a) The simplest approximation possible to account for scattering is simply to assume that scattering in the semiconducting nanotube can be treated exactly as scattering in a metallic nanotube using the concept of the effective mean free path embedded within the Landauer transmission probability, as discussed in Chapter 7. Further assume that the Landauer transmission probability is bias independent. Go ahead and incorporate this transmission probability in the integral expression for the drain current and derive the analytical diffusive $I-V$ formula.

(b) Let us improve on the approximation in part (a) by assuming that, in general, this transmission probability is at least dependent on the gate voltage, and this dependency can be expanded as a power series. For this improved case, re-derive the $I-V$ formula. (*Hint*: an analytical solution of the integral is possible).

8.3. Validity of the first-subband approximation for transistor $I-V$ characteristics.

In the development of the transistor characteristics of a nanotube FET with an intrinsic channel, we simply postulated that the first-subband largely determines current for practical VLSI technologies with sub-1 V power supply. Let us examine this using a simple model and quantitatively access the impact of higher subbands. For this exercise, the bandgap of the model CNT is taken to be 0.5 eV, the power supply is 0.5 V, $V_{FB} = 0$ V and feel free to employ the approximation $\varphi_s \approx V_G$ for mathematical convenience. We restrict the analysis to examine just the impact of the second subband, because if the second subband offers negligible contribution, then the higher subbands do not matter.

(a) Derive the analytical $I-V$ transistor relation including the first two subbands for a ballistic n-CNFET with ideal transparent contacts.
(b) For the model conditions stated above, how much larger (or smaller) is the two-subband ON current compared with just the first-subband I_{ON}?
(c) Redo part (b) but now take the power supply to be 0.8 V. Is the second subband still negligible?

8.4. Transmission probabilities for Schottky-barrier p-CNFETs.
The transmission probabilities for a Schottky-barrier n-CNFET are stated in Eqs. (8.47) and (8.48). Adapt these formulas for a Schottky-barrier p-CNFET.

8.5. Ambipolar suppression in CNFETs.
While the ambipolar characteristics of intrinsic CNFETs contacted by metals can be useful in optical applications, they are largely undesirable in conventional applications of transistors. Therefore, techniques to suppress ambipolar current are of fundamental interest. One straightforward proposal is to develop a complementary metal–oxide–semiconductor (CMOS)-like CNFET device structure (as in Figure 8.1c).[45] Let us define the model conditions for this device structure to aid us in obtaining qualitative insights. Consider both the intrinsic channel and the doped extension regions to be ballistic, the metal contacts are transparent, and $V_{FB} = 0$ V. We want to explore whether such a CMOS-like device structure can in fact suppress the ambipolar condition and, if it can, whether there are any obvious restrictions.

(a) Sketch the band diagram for the CMOS-like n-CNFET for both the OFF and ON states. Is ambipolar current suppressed in this CNFET? If ambipolar current is not suppressed, explain why. If it is suppressed, can we make a stronger statement and declare that ambipolar current is non-existent altogether?
(b) From inspection of the band diagram and for a given power supply V_{DD}, what are the specific restrictions (if any) on V_D and V_G in order

---

[45] Note that while Figure 8.1c is CMOS-like, it is not exactly the same as a conventional CMOS device. This is because, in a conventional CMOS device, the channel is also doped, as opposed to Figure 8.1c where the channel is intrinsic.

to eliminate ambipolar current? To aid in deducing an answer it might be worthwhile to consider $eV_{DD} < E_G$ and $eV_{DD} > E_G$.

(c) This device lends itself to the interesting device physics of source starvation or exhaustion.[25] Derive the transistor $I-V$ relations for $V_G < V_T$ and $V_G > V_T$, respectively. What is the ON current expression and dependence on V_G for both operating conditions?

(d) *Optional.* Propose a modified device structure that overcomes source starvation while retaining unipolar characteristics. (As in any other design problem, a variety of theoretically valid proposals might be possible.)

8.6. CNFETs with arrays of nanotubes.

We seek to gain some insight into the performance of ballistic ohmic contacted n-type CNFETs that contain multiple nanotubes in the channel. Ideally, all the nanotubes in the channel would be identical and semiconducting. However, in practice, the nanotubes generally have a distribution of diameters, say from 1 nm to 2 nm. To fully explore how the distribution in diameters affects the CNFET $I-V$ properties, it would be rigorously very satisfying to employ statistical methods to account for the diameter distribution and describe the composite $I-V$ properties. For the sake of time and information efficiency (insight per unit effort), let us consider a much simpler model with just a few nanotubes in the channel of the CNFET and assume operation in the quantum capacitance limit; after all, the purpose of this exercise is to guide you to some basic understanding, not necessarily to test your statistical abilities.

(a) The first insight we seek to acquire regards whether the distribution in diameters affects the switching properties of the CNFET. To address this question, let us focus exclusively on the sub-threshold slope as the metric of interest. The simplest model we can construct is a CNFET with two semiconducting nanotubes in the channel, one with a diameter: 1 nm and the other 1.5 nm. What is the sub-threshold slope S of this CNFET? Does the difference in diameters degrade the sub-threshold slope compared with an ideal CNFET where both nanotubes have identical diameters? (*Hint:* use the analytical ballistic $I-V$ theory in the quantum capacitance limit, and assume the DOPS parameter equals unity).

(b) The case of two different nanotubes in the channel is interesting; however, we would prefer the CNFET to contain much more than just two nanotubes in the channel. Let us proceed to make a broader generalization of the impact of diameter distribution on the sub-threshold slope of CNFETs. Consider $n = 10$ nanotubes with a uniform diameter distribution starting from 1 nm (i.e. 1 nm, 1.1 nm,... 1.9 nm). Using the exact analytical ballistic theory $I - V$ equation for the first subband in the quantum capacitance limit, plot the sub-threshold slope (S versus V_{gs}) for the

CNFET with n diameters of semiconducting nanotubes in the channel at room temperature with $V_{DD} = 1$ V and V_{gs} from 0 to 0.3 V. An intriguing question presents itself: Why is it that the distribution in diameters does or does not degrade the sub-threshold slope?

(c) Having all semiconducting nanotubes in the CNFET is currently a dream driving many researchers in the field. Presently there exists a probability of having metallic nanotubes (typically, one-third of CNTs are metallic). Even if a technique were developed today to synthesize semiconducting nanotubes, nature teaches us that the yield of enjoying exactly 100 % semiconducting CNTs is unlikely. What is more likely is that there will exist a finite (albeit however small) probability of metallic nanotubes. Inevitably, the natural question that arises is: What yield of semiconducting nanotubes is good enough for practical devices? Let us again focus on the sub-threshold slope to guide us in addressing this question. Consider a CNFET with n identical semiconducting nanotubes and one ideal ballistic metallic nanotube in the channel. Today's CMOS devices typically have a sub-threshold slope of about 80–100 mV/decade. How large should n be in order to achieve $S \sim 85$ mV/decade and what should the corresponding probability of growing semiconducting CNTs be? (*Hint*: employ the exact analytical ballistic theory $I-V$ equation for the first subband with S evaluated at $V_D = V_{DD} = 0.5$ V, and $eV_{gs} = E_g/2$ for analytical simplicity and $\alpha = 1$.)

9 Applications of carbon nanotubes

The unbeaten path is where discoveries of great ideas can be found.

9.1 Introduction

In contemporary fundamental and applied science research, the potential applications and the perceived broader impacts are undoubtedly the primary drivers for expanding the research enterprise. This has certainly been the case for nanotube research. The unique unprecedented properties of CNT, such as their perfect tubular structure, outstanding electrical and thermal conductance, tunable optical properties, and superior mechanical strength and stiffness, have generated great excitement, leading to the pursuit of both fundamental insights of the beauty of nature in reduced dimensions of condensed matter, and the novel applications and technological breakthroughs that can be developed. In essence, the exploration of nanotubes (and other nanomaterials) is to learn about their nature and their interaction with fields and matter that will allow us to synthesize CNTs, design devices, and develop unique materials for next-generation transformative products. This endeavor has brought together many parties across several boundaries of knowledge, from nanomedicine to nanoscience to nanotechnology.

To put CNT in a broader perspective, over the last decade, nanotube applied research and development in academic and industrial laboratories across the world has enjoyed a substantial rise, reflecting a rise in the deeper understanding of the material. Figure 9.1 shows the increase in CNT patent applications and patents issued in the United States. It is an indicator of the growing effort to employ nanotubes in innovative applications. Invariably, many of the applications of CNTs take advantage of their inherent nanoscale dimension, large surface-to-volume ratio, and unique combination of electrical, optical, thermal, and structural properties.

This chapter is a survey of applications of CNTs. We do not attempt to predict the future, but merely give an overview of some of the growing applications of CNTs. Owing to the limited context of the present textbook, it is not possible to summarize every promising use of nanotubes. As such, we will simply highlight some of the prominent applications and provide relevant references along the way

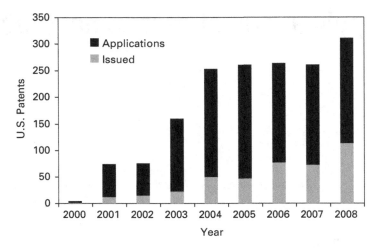

Fig. 9.1 United States patents issued and patent applications containing the phrase "carbon nanotube" in the patent abstract. In a sense, this is an indicator of the growing interest in applications of CNTs.

for the interested reader to dig deeper.[1] Additionally, there are several nanotube edited volumes offered by the technical publishing press which are useful for further discussion.[2] In this chapter, nanotubes will generally refer to both the single-wall and the multi-wall flavors.

9.2   Chemical sensors and biosensors

Nanomaterials are ideal materials for chemical sensors and biosensors because of their reduced dimensionality and large surface-to-volume ratio. Ideally, every atom is physically accessible under any ambient condition. The direct equal access to every atom and the ability to perturb each atom and measure the perturbations electrically or optically are the essential attributes that make CNTs and graphene highly attractive for sensor applications. Additionally, the structural stability and tunable electrical/optical properties of CNTs greatly encourage the exploration and development of a wide range of chemical, molecular and biological sensors that can take advantage of the century-old wealth of knowledge about carbon structures formally embodied in organic chemistry.

[1] Owing to the large number of references per page, the listing of references in this chapter will depart from the style of the previous chapters. All published references will be listed conveniently at the end of the chapter. Also, the reader will find the references contained in the cited references invaluable for comprehensive further studies of the particular CNT application.

[2] See M. Meyyappan (ed), *Carbon Nanotubes Science and Applications* (CRC Press, 2005); and O. Brand, G. K. Fedder, C. Hierold, J. G. Korvink and O. Tabata (editors), *Carbon Nanotube Devices: Properties, Modeling, Integration and Applications* (Wiley-VCH, 2008).

Even though there are a variety of techniques to employ CNTs as sensors, the main operating principle can be summarized as follows: pristine, high-quality nanotubes are chemically treated by applying a suitable coating or attaching a specific functional group to enhance their sensitivity and selectivity to the species or interactions of interest, or to adapt them to the host environment (e.g. for biosensors in solution phase). The subsequent changes in the CNT electrical current/conductance or in their optical emission provide information regarding the interaction with or presence of the target species. Using this principle, many types of nanotube-based chemical and gas sensors have been demonstrated [1–7]. In other cases, CNTs are used as an electrode for capacitance-based sensors [8], or are employed to generate high electric fields at their tips to ionize a gas [9]. The resulting gas ionization or breakdown voltage informs on the nature of the ambient gas. Moreover, multi-functional sensors with CNTs have been investigated for electronic nose (e-nose) applications [10]. As with non-nanotube sensors, enhanced selectivity and increased sensitivity are seemingly never-ending themes of sensor research and development. Currently, CNT-based sensors have been shown to offer parts per billion types of sensitivity for select gases [2, 3]. The nanotube-based sensors are typically constructed out of two-terminal CNT devices or three-terminal transistor-like structures.

For biosensors, the small nanoscale dimension of CNTs is ideal for interacting with bio-molecules. Figure 9.2 is an example of a nanotube biosensor for optically detecting toxins. Typically, the nanotubes are treated (for example, enzyme coated or DNA wrapped, as shown in Figure 9.2) for a variety of reasons, including the need to be compatible with the biological environment or to monitor/sense specific proteins that convey information about biological events and processes [4]. However, there have been mixed results regarding the biocompatibility and toxicity of nanotubes under *in-vivo* conditions [11–13] which has generated considerable debate in the scientific and popular media. Obviously, this is a matter of concern on the minds of both researchers and policymakers alike, especially with regard to clinical applications and public health. In fact, a book with a catchy title exploring the safety of nanotubes has recently been published,[3] and is useful reading for a broader assessment on the state of affairs.

As nanotube sensors mature, there is a growing trend to integrate them with silicon electronics for calibrated, improved, and robust sensor control and performance [14, 15]. At the same time, modeling of sensor behavior is important for the analysis and design of high-performance CNT-based chemical sensors and biosensors [16]. On a related note for bio applications, CNTs are also actively being explored for cancer treatment and as a nano-vehicle for drug delivery [17, 18].

---

[3] S. Fiorito, *Carbon Nanotubes: Angels or Demons?* (Pan Stanford Publishing, 2008).

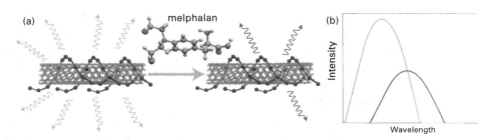

Fig. 9.2   Detecting toxins with CNTs. (a) A strand of DNA wrapped around a (6, 5) nanotube has photoluminescence at a near-infrared wavelength. In the presence of melphalan, a chemotherapy drug, the emission moves to a longer wavelength (i.e. red-shifted). (b) Graph of the initial and reduced (red-shifted) photoluminescence intensity of a DNA-wrapped CNT in the presence of melphalan. Reprinted by permission from Macmillan Publishers Ltd: T. D. Krauss [7], copyright (2009).

9.3   Probe tips for scanning probe microscopy

Scanning probe microscopy in its many variants, including AFM and STM, has become a necessary and routine imaging technique, leading to better understanding of surface topography, surface physics, and surface chemistry of materials and matter since its invention [19]. The deeper understanding provided by the use of scanning probes continues to enable advancements in metrology, materials science, nanoelectronics and nanolithography, condensed matter physics, physical chemistry, and structural biology [20].

An essential component of all scanning probe systems is the probe tip, which is attached to the end of a cantilever and moved about the sample surface in a precise manner in order to measure the surface properties of interest. The lateral dimension of the probe tip determines the spatial (lateral) resolution, while its vertical length determines the ability to image deep trenches accurately. Naturally, the slender cylindrical structure of CNTs with diameters on the order of nanometers and lengths on the order of micrometers coupled with their axial stiffness makes CNTs a near-ideal probe tip [21–23], with significantly improved resolution beyond what can be achieved with standard silicon probe tips (see Figure 9.3). In addition, the ability to functionalize the end of the nanotube tip with chemical groups adds to the portfolio of imaging capabilities, to include high-resolution imaging of surface molecular interactions with potential applications in many areas of chemistry and biology [24].

The cylindrical nature of CNTs imposes a trade-off between the lateral resolution and the depth (trench) resolution. Though CNTs are very stiff, their length nonetheless needs to be commensurate with their diameter in order to achieve robust operation over repeated use. This implies that for maximum lateral resolution with a nanotube of about 1nm diameter, the corresponding length needs to

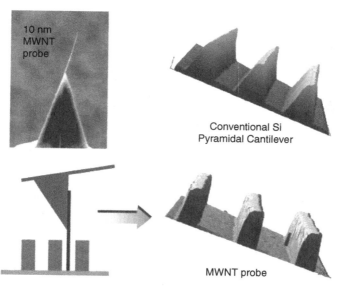

Fig. 9.3 Comparison of AFM images obtained with a standard silicon probe tip and that obtained using a multi-wall CNT (MWNT) showing a finer lateral and vertical resolution. Reprinted with permission from [22] and [23] (copyright 2007 American Chemical Society).

be short enough ($\sim$10 nm) to maintain sufficient stiffness for reliable usage, while imaging of micrometer deep trenches will nominally require CNTs with diameters greater than 10 nm [20]. Alternatively, thin nanotubes can be coated to enhance their rigidity and also their length [25]. The coating can be removed at the very tip of the nanotube, essentially modifying the cylindrical structure of the tip to a somewhat conical structure, thereby providing both fine lateral and depth resolution on flat and corrugated surfaces.

There are variety of methods that have been developed to manufacture nanotube integrated probe tips. In fact, several probe manufacturers already offer nanotube probe tips as part of their catalog, although at a relatively premium price. Continued progress in controllable large-scale nanotube synthesis on microfabricated cantilevers will go a long way in making nanotube probes more affordable and encourage their routine usage for scanning probe microscopy.

9.4 Nano-electromechanical systems (NEMS)

Nano-electromechanical systems (NEMS) are devices that integrate electrical and mechanical functionality at nanoscale dimensions. It is fair to say that an essential component of many NEMS devices is the cantilever beam, which is illustrated with a CNT in Figure 9.4. The cantilever typically serves as a physical sensor, where its position or motion is a function of its environmental conditions, or as an actuator

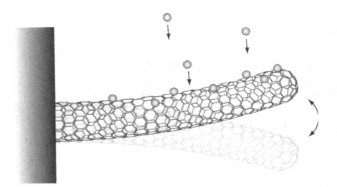

Fig. 9.4   An atomic-resolution mass sensor using a CNT cantilever [32]. (Courtesy of Zettl Research Group, Lawrence Berkeley National Laboratory and University of California at Berkeley.)

that can be electrically controlled. Naturally, CNTs are suitable as cantilevers for the design of NEMS owing to their slender structure, high directional stiffness, low mass, chemical inertness, attractive electrical and thermal properties, and the ability to accommodate relatively very large strains without breaking.

Nanotube-based NEMS have shown promising performance in experimental systems, including sensitive pressure sensors [26], nanotweezers [27], high-speed nanorelays [28], non-volatile memory [29], tunable oscillators [30], and a multifunctional single nanotube-NEMS radio [31]. Recently, an atomic-resolution mass sensor using a CNT cantilever, as shown in Figure 9.4 was demonstrated [32]. Additionally, nanotube cantilevers are candidates for the rising field of complementary NEMS relays, which are currently the only known nanoscale switching devices that offer virtually zero leakage current.

Sustained exploration of nanotube-based NEMS will go a long way in optimizing the device performance and developing the technology to commercial maturity.

9.5   Field emission of electrons

Electron guns based on field emission of electrons from the tips of vertically aligned CNTs was one of the earliest properties of CNTs investigated [33, 34], generating a wave of industrial and scientific explorations into various potential applications, including vacuum electronics, cold cathodes, electron microscopes [35], field emission displays [36],[4] and X-ray sources for medical applications [37, 38]. Any system that employs an electron source could potentially use a nanotube-based

---

[4] Samsung was successful in demonstrating a CNT-based electronic display; however, it was not released as a product because it was apparently relatively too expensive.

electron emission device. Field emission of electrons works by the application of a sufficiently high electric field to the tip of the nanotube to emit electrons. The lean mechanically firm structure of nanotubes combined with their high thermal conductivity and chemical stability are particularly desirable qualities of a field emitter; in addition, their small radius affords high field-enhancement factors. The nanotubes are typically grown on a conductive substrate in the vertical direction by CVD and patterned by lithography for size control of the CNT emitters, as shown in Figure 9.5a. Continuous electron emission is obtained by applying a sufficiently high electric field (usually in high vacuum) between the anode (the positive top plate some distance away from the CNTs) and the cathode (the CNT conductive substrate).

The theory of electron emission from metals is commonly described by Fowler–Nordheim tunneling [40], or a modified version [41]. Often it is assumed that the electron emission process from a nanotube is like that of a metallic sharp tip, an assumption that has been validated in several experiments [39, 42]. Figure 9.5c shows the current density for a high-performance CNT field emission device with

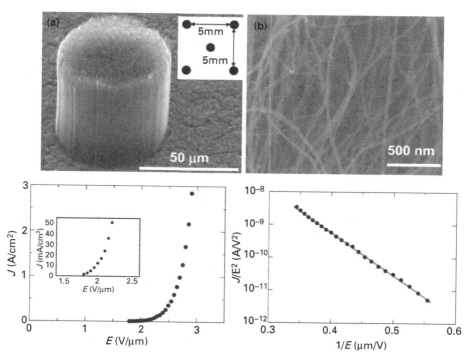

Fig. 9.5 Characteristics of a CNT field emission device. (a) Electron microscope image of bundles of nanotubes with a diameter of 50 μm and height of 70 μm. (b) Enlarged image of the bundle. (c) Emission current density plot. (d) Fowler–Nordheim log-linear plot. Images and data reprinted with permission from Fujii *et al.* [39]. Copyright (2007), American Institute of Physics.

current densities of $10\,\text{mA cm}^{-2}$ at a low-threshold electric field of $2\,\text{V}\,\mu\text{m}^{-1}$, and exceeding $1\,\text{A cm}^{-2}$ for fields $<3\,\text{V}\,\mu\text{m}^{-1}$ which is competitive compared with conventional field emitters. Figure 9.5d is the Fowler–Nordheim log-linear plot, where the straight line profile is characteristic of Fowler–Nordheim tunneling. Besides a low-threshold field and high current density, an additional performance parameter for practical electron emitters includes long-term current stability, which nanotubes have been demonstrated to achieve [33, 36, 39]. Applications of nanotube field emitters appear to be progressing favorably [37, 38, 43].

9.6   Integrated electronics on flexible substrates

An emerging technology over the past $\sim$15 years has been the integration of transistors and passive components on flexible plastic substrates for large-area electronics that are conformal, lightweight, and shock resistant, all at a relatively low cost [44, 45]. The wide range of applications is constantly evolving to include: large-area flexible displays, electronic paper, low-cost photovoltaics, wearable sensor electronics, and structure-conforming electronics for mobile systems or moving vehicles. These types of integrated electronics are difficult or impossible to achieve on rigid substrates such as silicon wafers. Moreover, implementations of large-area electronics would be extremely cost prohibitive on semiconductor substrates with conventional semiconductor fabrication techniques using vacuum deposition and etching, which are geared to produce micro-scale electronic chips.

Fabrication of passive components, such as inductors, capacitors, and interconnect wires, on a variety of plastic and ceramic substrates using inexpensive printing techniques has existed for many decades, and one popular variant is the printed circuit board. The main challenge in the evolution of printed electronics is the monolithic integration of a suitable semiconductor with acceptable performance to serve as the active component. The printed, or in many cases evaporated, semiconductor is commonly an organic semiconductor, such as pentacene [46, 47].

Recently, aligned arrays or networks of interconnected nanotubes have been identified as suitable semiconductors for integration on flexible substrates. The goal is to access the intrinsic performance benefits of CNTs, such as high mobilities, high current densities, high optical transmittance, and robust mechanical properties, on a flexible platform. To integrate a transistor on a plastic substrate requires deposition techniques for gate electrodes, gate dielectrics, semiconducting material, and source/drain electrodes. Use of conventional vacuum deposition techniques leads to evaporated conductors, dielectrics, and a suitable organic semiconductor such as pentacene. Printing techniques employ solution-processed conductors, dielectrics, and semiconductors. By transferring CNTs grown on an oxidized silicon wafer

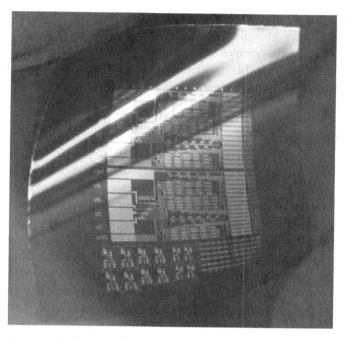

Fig. 9.6   Photograph of an assortment of CNT transistors and circuits on a thin plastic (polyimide) sheet. Reprinted by permission from Macmillan Publishers Ltd: Nanotechnology [45], copyright (2008).

to a plastic substrate (e.g. polyimide), and using photolithography with conventional deposition techniques, high-performance nanotube transistors and circuits have been realized on a flexible platform, as shown in Figure 9.6. The promising performance of experimental CNFETs on flexible substrates (nominal mobility $\sim$70 cm^2 V^{-1} s^{-1}, transconductance $\sim$0.12 μS μm^{-1}, and sub-threshold slope $\sim$200 mV/decade for channel lengths $\geq$50 μm, operating at voltages <5 V [45]) expands the application possibilities of flexible integrated circuits to include high-frequency communication systems at low cost. For example, wireless displays and sensors, active antenna systems [44], and RF identification (RFID) tags [46, 47] become realistic systems that can be fully implemented on plastic substrates as device mobilities and speed increases.[5] By a similar token, increasing mobilities relax the constraints and tolerances on the critical dimensions and multi-level alignment to reduce the cost of printing techniques such as roll-to-roll fabrication or graphic arts printing technologies.

---

[5] Device mobility is a key metric for integrated transistors on plastics and in a sense defines the application space. Currently, device mobilities of the order of 0.01–10 cm^2 V^{-1} s^{-1} are fairly common for organic semiconductors, with values on the lower end being more prevalent.

Contemporary research is geared towards further (reproducible) increases in device mobilities while developing the technology infrastructure, such as well-characterized process recipes, standardized design guidelines, and device/circuit modeling, for high-yield fabrication of integrated electronics on flexible or low-cost substrates for a wide range of applications.

9.7   Hydrogen storage

The increasing awareness of the adverse effects of environmental pollution has led to great interest and efforts to develop so-called *green technologies*.[6] Undoubtedly, one of the major sources of pollution comes about as by-products from the burning of crude oil or similar hydrocarbons in order to provide energy to move vehicles for the transportation of people and goods. In this regard, an alternative cleaner method of providing energy for transportation is a much sought after technology with global ramifications. A rising option is to employ the electrical energy released in the oxidation of hydrogen to power moving vehicles. There are several reasons that motivate the choice of hydrogen, including [48, 49]:

(i)   its high energy content (one accessible electron for electricity per proton);
(ii) hydrogen is the most abundant element in the world (in the form of water);
(iii) its by-product or oxidation product associated with energy generation is water, which is environmentally benign.

Though hydrogen is the most abundant element in the world, it mostly exists in a bound form in water. However, for practical use it needs to be in a free form. One vision to produce hydrogen in a clean manner is to harvest the energy present in sunlight for the electrolysis of water into hydrogen and oxygen in a photoelectrochemical cell. The process is commonly known as "water splitting." The next challenge in a hydrogen-powered transportation system involves the distribution and storage of hydrogen.

The pioneering work of Dillon *et al.* [50] brought to light the potential of nanotubes as a medium for the storage of hydrogen, say on board a vehicle. Single-wall CNTs are particularly attractive for hydrogen storage because of their light mass and fully exposed atomic structure, which implies that all the carbon atoms can participate in adsorbing hydrogen. Additionally, their empty interior can serve as some kind of nano-container for hydrogen. Figure 9.7 shows some representative ways hydrogen can be stored on the exterior and interior of CNTs. Note that the nanotubes generally expand in order to accommodate hydrogen. A metric used to

---

[6] Green technology is an environmentally friendly technology meaning a technology that in a very broad sense is gentle on the environment. At the very core is the practice of ideas embracing sustainability, renewability, conservation, and energy efficiency. In some English dictionaries (such as Oxford and Microsoft Word), green has become a synonym for the environment.

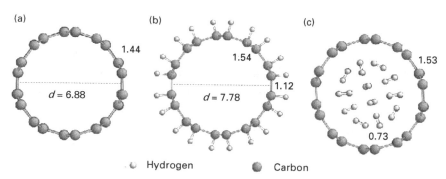

Fig. 9.7 Illustrations of hydrogen adsorption on a (5, 5) single-wall nanotube. (a) A pristine CNT. (b) Hydrogen atoms adsorbed on the exterior of the CNT surface ($\theta = 1$). (c) Adsorption of hydrogen atoms in the interior of the nanotube ($\theta = 1.2$). The bond lengths and diameters d are in angströms. Reprinted with permission from Lee and Lee [52]. Copyright (2000), American Institute of Physics.

quantify the hydrogen storage capacity C_H is the percentage normalized weight, given by

$$C_H = \frac{\theta A_H}{\theta A_H + A_C} \text{ (weight\%)} \qquad (9.1)$$

where A_H and A_C are the atomic weights of hydrogen and carbon, which are 1 g and 12 g respectively, and θ is the ratio of hydrogen atoms to carbon atoms. For the exterior adsorption of a monolayer of hydrogen, the maximum hydrogen storage capacity ($\theta = 1$) is ~7.7 %, normally written in the form 7.7 wt%. This value is of significance because it exceeds the U.S. DOE *ultimate* gravimetric capacity target.[7] Indeed, experiments on the exterior hydrogenation of CNTs have achieved storage capacities greater than 7 wt% that are stable at room temperature [51]. Theoretically, the maximum stable intercalation of hydrogen in the interior of the nanotubes is predicted to be around $\theta \sim 2.0$, corresponding to about 14.3 wt% [52]. This suggests that the use of single-wall nanotubes for hydrogen storage is a promising realistic technology. However, there are significant challenges to be overcome in meeting the U.S. DOE volumetric capacities in a practical system.[7] Research is ongoing, and the considerable rewards waiting at the end of the journey continue to inspire further explorations.

---

[7] The U.S. DOE (Department of Energy) National Hydrogen Storage Project outlines a roadmap of requirements for on-board hydrogen storage. The 2015 targets for gravimetric and volumetric capacities are 5.5 wt% and 40 g L^{-1} respectively. The *ultimate* targets are 7.5 wt% and 70 g L^{-1}. See http://www1.eere.energy.gov/hydrogenandfuelcells/storage/current_technology.html (accessed December 2009).

9.8   Composites

Composite materials loaded with carbon matter were one of the initial applications of tubular carbon in modern history, dating back to the late 19th century, where carbon fibers were used as high-temperature conductive filaments in the early development of the lighting industry.[8] Later, in the mid 1900s, there was a resurgence of interest in carbon fiber for high-temperature lightweight composites for aircraft and satellites, pioneered by the work of Roger Bacon, a physicist working on carbon fiber synthesis. The same attractive thermal, mechanical, and electrical properties of carbon fibers are even more pronounced in CNTs, and this is the reason for the current renaissance in CNT composites. Currently, CNTs are the strongest, most resilient material known, and at the same time offer superior electrical and thermal conductivities exceeding those of silver and diamond respectively [53, 54]. The primary motivation is to confer the outstanding physical properties of CNTs to the composite material to achieve overall enhanced performance that can be tailored for the specific application at hand. In many cases, the composite is typically processed or loaded with 5–10 wt% of CNTs.

Today, the wide range of applications has evolved beyond the traditional base and is more or less limited by the imagination at the moment. Obvious applications continue to include lightweight, high-temperature composites with strengths that are routinely greater than steel, and which in some cases can be synthesized to be comparable to that of the resilient spider silk [54]. Novel applications, such as sensors, supercapacitors [55], conductive polymers for photovoltaics [56], nanotube-reinforced copper for high-reliability interconnects in nanoscale technology [57], CNT composite fibers for electronic textiles [54], CNT-coated electrodes for probing neurons in the brain [58], and bio-compatible materials for scaffolding in tissue engineering [59], are now included in the application catalog. Quite frankly, there is no shortage of innovative and unusual ideas. The space elevator is perhaps the most widely reported example of a very inspired use of nanotube composites.[9] It is noteworthy that consumer products reinforced with CNTs, such as premium-quality tennis rackets and bikes, are already available in the marketplace.[10]

---

[8] For example, the development of the first successful incandescent light bulbs by Joseph Swan and later by Thomas Edison used carbon fibers as the filament.

[9] Wikipedia has an interesting article on nanotubes and the space elevator idea. See http://en.wikipedia.org/wiki/space_elevator.

[10] The racquet manufacturer Babolat sells CNT-reinforced tennis racquets. Also, the Phonak cycling team used nanotube-reinforced bikes from BMC for the 2006 Tour de France competition.

9.9   References

[1] P. G. Collins, K. Bradley, M. Ishigami and A. Zettl, Extreme oxygen sensitivity of electronic properties of carbon nanotubes. *Science*, **287** (2000) 1801–4.

[2] P. Qi, O. Vermesh, M. Grecu, A. Javey, Q. Wang, H. Dai, S. Peng and K. J. Cho, Toward large arrays of multiplex functionalized carbon nanotube sensors for highly sensitive and selective molecular detection. *Nano Lett.*, **3** (2003) 347–51.

[3] J. Li, Y. Lu, Q. Ye, M. Cinke, J. Han and M. Meyyappan, Carbon nanotube sensors for gas and organic vapor detection. *Nano Lett.*, **3** (2003) 929–33.

[4] K. Besteman, J. O. Lee, F. G. M. Wiertz, H. A. Heering and C. Dekker, Enzyme-coated carbon nanotubes as single-molecule biosensors. *Nano Lett.*, **3** (2003) 727–30.

[5] J. Wang, carbon-nanotube based electrochemical biosensors: a review. *Electroanalysis*, **17** (2005) 7–14.

[6] P. W. Barone, S. Baik, D. A. Heller and M. S. Strano, Near-infrared optical sensors based on single-walled carbon nanotubes. *Nat. Mater.*, **4** (2005) 86–92.

[7] T. D. Krauss, Biosensors: nanotubes light up cells. *Nat. Nanotechnol.*, **4**, (2009) 85–6.

[8] E. S. Snow, F. K. Perkins, E. J. Houser, S. C. Badescu and T. L. Reinecke, Chemical detection with a single-walled carbon nanotube capacitor. *Science*, **307** (2005) 1942–45.

[9] A. Modi, N. Koratkar, E. Lass, B. Wei and P. M. Ajayan, Miniaturized gas ionization sensors using carbon nanotubes. *Nature*, **424** (2003) 171–4.

[10] P.-C. Chen, F. N. Ishikawa, H.-K. Chang, K. Ryu and C. Zhou, A nanoelectronic nose: a hybrid nanowire/carbon nanotube sensor array with integrated micromachined hotplates for sensitive gas discrimination. *Nanotechnology*, **20** (2009) pp 125503.

[11] S. K. Smart, A. I. Cassady, G. Q. Lu and D. J. Martin, The biocompatibility of carbon nanotubes. *Carbon*, **44** (2006) 1034–47.

[12] C. A. Poland, R. Duffin, I. Kinloch, A. Maynard, W. A. H. Wallace, A. Seaton, V. Stone, S. Brown, W. MacNee and K. Donaldson, Carbon nanotubes introduced into the abdominal cavity of mice show asbestos-like pathogenicity in a pilot study. *Nat. Nanotechnol.*, **3** (2008) 423–8.

[13] Z. Liu, C. Davis, W. Cai, L. He, X. Chen and H. Dai, Circulation and long-term fate of functionalized, biocompatible single-walled carbon nanotubes in mice probed by Raman spectroscopy. *Proc. Nat. Acad. of Sci.*, **105** (2008) 1410.

[14] T. S. Cho, K. J. Lee, J. Kong and A. P. Chandrakasan, A low power carbon nanotube chemical sensor system. In *IEEE Custom Integrated Circuits Conf.*, 2007, 181–4.

[15] D. Akinwande, S. Yasuda, B. Paul, S. Fujita, G. Close and H.-S. P. Wong, Monolithic integration of CMOS VLSI and carbon nanotubes for hybrid nanotechnology applications. *IEEE Trans. Nanotechnol.*, **7** (2008) 636–9.

[16] J. Deng, K. Ghosh and H.-S. P. Wong, Modeling carbon nanotube sensors. *Sensors J., IEEE*, **7** (2007) 1356–7.

[17] A. Bianco, K. Kostarelos and M. Prato, Applications of carbon nanotubes in drug delivery. *Curr. Opin. Chem. Biol.*, **9** (2005) 674–9.

[18] Z. Liu, K. Chen, C. Davis, S. Sherlock, Q. Cao, X. Chen and H. Dai, Drug delivery with carbon nanotubes for *in vivo* cancer treatment. *Cancer Res.*, **68** (2008) 6652–60.

[19] G. Binnig, H. Rohrer, C. Gerber and E. Weibel, Surface studies by scanning tunneling microscopy. *Phys. Rev. Lett.*, **49** (1982) 57.

[20] N. R. Wilson and J. V. Macpherson, Carbon nanotube tips for atomic force microscopy. *Nat. Nanotechnol.*, **4** (2009) 483–91.

[21] H. Dai, J. H. Hafner, A. G. Rinzler, D. T. Colbert and R. E. Smalley, Nanotubes as nanoprobes in scanning probe microscopy. *Nature*, **384** (1996) 147–50.

[22] C. V. Nguyen, K.-J. Chao, R. M. D. Stevens, L. Delzeit, A. Cassell, J. Han and M. Meyyappan, Carbon nanotube tip probes: stability and lateral resolution in scanning probe microscopy and application to surface science in semiconductors. *Nanotechnology*, **12** (2001) 363–67.

[23] M. Valcarcel, S. Cardenas and B. M. Simonet, Role of carbon nanotubes in analytical science. *Anal. Chem.*, **79** (2007) 4788–97.

[24] S. S. Wong, E. Joselevich, A. T. Woolley, C. L. Cheung and C. M. Lieber, Covalently functionalized nanotubes as nanometre-sized probes in chemistry and biology. *Nature*, **394** (1998) 52–5.

[25] A. Patil, J. Sippel, G. W. Martin and A. G. Rinzler, Enhanced functionality of nanotube atomic force microscopy tips by polymer coating. *Nano. Lett.*, **4** (2004) 303–8.

[26] C. Stampfer, T. Helbling, D. Obergfell, B. Schoberle, M. K. Tripp, A. Jungen, S. Roth, V. M. Bright and C. Hierold, Fabrication of single-walled carbon-nanotube-based pressure sensors. *Nano. Lett.*, **6** (2006) 233–7.

[27] P. Kim and C. M. Lieber, Nanotube Nanotweezers. *Science*, **286** (1999) 2148–50.

[28] A. B. Kaul, E. W. Wong, L. Epp and B. D. Hunt, Electromechanical carbon nanotube switches for high-frequency applications. *Nano. Lett.*, **6** (2006) 942–7.

[29] T. Rueckes, K. Kim, E. Joselevich, G. Y. Tseng, C.-L. Cheung and C. M. Lieber, Carbon nanotube-based nonvolatile random access memory for molecular computing. *Science*, **289** (2000) 94–7.

[30] V. Sazonova, Y. Yaish, H. Ustunel, D. Roundy, T. A. Arias and P. L. McEuen, A tunable carbon nanotube electromechanical oscillator. *Nature*, **431** (2004) 284–7.

[31] K. Jensen, J. Weldon, H. Garcia and A. Zettl, Nanotube Radio. *Nano. Lett.*, **7** (2007) 3508–11.

[32] K. Jensen, K. Kim and A. Zettl, An atomic-resolution nanomechanical mass sensor. *Nat. Nanotechnol.*, **3** (2008) 533–7.

[33] W. A. de Heer, A. Châtelain and D. Ugarte, A Carbon Nanotube Field-Emission Electron Source. *Science*, **270** (1995) 1179–80.

[34] A. G. Rinzler, J. H. Hafner, P. Nikolaev, P. Nordlander, D. T. Colbert, R. E. Smalley, L. Lou, S. G. Kim and D. Tomanek, Unraveling nanotubes: field emission from an atomic wire. *Science*, **269** (1995) 1550–3.

[35] N. de Jonge, Y. Lamy, K. Schoots and T. H. Oosterkamp, High brightness electron beam from a multi-walled carbon nanotube. *Nature*, **420** (2002) pp. 393–5.

[36] W. B. Choi, D. S. Chung, J. H. Kang, H. Y. Kim, Y. W. Jin, I. T. Han, Y. H. Lee, J. E. Jung, N. S. Lee, G. S. Park and J. M. Kim, Fully sealed, high-brightness carbon-nanotube field-emission display. *Appl. Phys. Lett.*, **75** (1999) 3129–31.

[37] G. Z. Yue, Q. Qiu, B. Gao, Y. Cheng, J. Zhang, H. Shimoda, S. Chang, J. P. Lu and O. Zhou, Generation of continuous and pulsed diagnostic imaging x-ray radiation using a carbon-nanotube-based field-emission cathode. *Appl. Phys. Lett.*, **81** (2002) 355–7.

[38] G. Cao, Y. Z. Lee, R. Peng, Z. Liu, R. Rajaram, X. Calderon-Colon, L. An, P. Wang, T. Phan and S. Sultana, A dynamic micro-CT scanner based on a carbon nanotube field emission X-ray source. *Phys. Med. Biol.*, **54** (2009) 2323–40.

[39] S. Fujii, S.-I. Honda, H. Machida, H. Kawai, K. Ishida, M. Katayama, H. Furuta, T. Hirao and K. Oura, Efficient field emission from an individual aligned carbon nanotube bundle enhanced by edge effect. *Appl. Phys. Lett.*, **90** (2007) 153108.

[40] R. H. Fowler and L. Nordheim, Electron emission in intense electric fields. *Proc. R. Soc. of London Ser. A: Contain. Pap. Math. Phys. Char.*, **119** (1928) 173–81.

[41] R. G. Forbes, Simple good approximations for the special elliptic functions in standard Fowler–Nordheim tunneling theory for a Schottky–Nordheim barrier. *Appl. Phys. Lett.*, **89** (2006) 113122.

[42] N. de Jonge, M. Allioux, M. Doytcheva, M. Kaiser, K. B. K. Teo, R. G. Lacerda and W. I. Milne, Characterization of the field emission properties of individual thin carbon nanotubes. *Appl. Phys. Lett.*, **85** (2004) 1607–9.

[43] L. F. Velasquez-Garcia and A. I. Akinwande, Fabrication of large arrays of high-aspect-ratio single-crystal silicon columns with isolated vertically aligned multi-walled carbon nanotube tips. *Nanotechnology*, **19** (2008) 405305.

[44] R. H. Reuss, B. R. Chalamala, A. Moussessian, M. G. Kane, A. Kumar, D. C. Zhang, J. A. Rogers, M. Hatalis, D. Temple, G. Moddel, B. J. Eliasson, M. J. Estes, J. Kunze, E. S. Handy, E. S. Harmon, D. B. Salzman, J. M. Woodall, M. A. Alam, J. Y. Murthy, S. C. Jacobsen, M. Olivier, D. Markus, P. M. Campbell and E. Snow, Macroelectronics: perspectives on technology and applications. *Proc. IEEE*, **93** (2005) 1239–56.

[45] Q. Cao, H.-S. Kim, N. Pimparkar, J. P. Kulkarni, C. Wang, M. Shim, K. Roy, M. A. Alam and J. A. Rogers, Medium-scale carbon nanotube thin-film integrated circuits on flexible plastic substrates. *Nature*, **454** (2008) 495–500.

[46] A. Dodabalapur, Organic and polymer transistors for electronics. *Mater. Today*, **9** (2006) 24–30.

[47] M. Chason, P. W. Brazis, J. Zhang, K. Kalyanasundaram and D. R. Gamota, Printed organic semiconducting devices. *Proc. IEEE*, **93** (2005) 1348–56.

[48] L. Schlapbach and A. Zuttel, Hydrogen-storage materials for mobile applications. *Nature*, **414** (2001) 353–8.

[49] R. Ströbel, J. Garche, P. T. Moseley, L. Jörissen and G. Wolf, Hydrogen storage by carbon materials. *J. Power Sources*, **159** (2006) 781–801.

[50] A. C. Dillon, K. M. Jones, T. A. Bekkedahl, C. H. Kiang, D. S. Bethune and M. J. Heben, Storage of hydrogen in single-walled carbon nanotubes. *Nature*, **386** (1997) 377–9.

[51] A. Nikitin, X. Li, Z. Zhang, H. Ogasawara, H. Dai and A. Nilsson, Hydrogen storage in carbon nanotubes through the formation of stable C–H bonds. *Nano. Lett.*, **8** (2007) 162–7.

[52] S. M. Lee and Y. H. Lee, Hydrogen storage in single-walled carbon nanotubes. *Appl. Phys. Lett.*, **76** (2000) 2877–9.

[53] E. T. Thostenson, Z. Ren and T. W. Chou, Advances in the science and technology of carbon nanotubes and their composites: a review. *Compos. Sci. Technol.*, **61** (2001) 1899–912.

[54] A. B. Dalton, S. Collins, E. Munoz, J. M. Razal, V. H. Ebron, J. P. Ferraris, J. N. Coleman, B. G. Kim and R. H. Baughman, Super-tough carbon-nanotube fibres. *Nature*, **423** (2003) 703.

[55] E. Frackowiak, Carbon materials for supercapacitor application. *Phys. Chem. Chem. Phys.*, **9** (2007) 1774–85.

[56] E. Kymakis and G. A. J. Amaratunga, Single-wall carbon nanotube/conjugated polymer photovoltaic devices. *Appl. Phys. Lett.*, **80** (2002) 112–4.

[57] C. Yang and P. Chan, High electromigration-resistant copper/carbon nanotube composite for interconnect application. Presented at Electron Devices Meeting, 2008. IEDM 2008. IEEE International, 2008.

[58] E. W. Keefer, B. R. Botterman, M. I. Romero, A. F. Rossi and G. W. Gross, Carbon nanotube coating improves neuronal recordings. *Nat. Nanotechnol.*, **3** (2008) 434–9.

[59] A. M. Rebecca, F. L. Brendan, V. Gunaranjan, M. A. Pulickel and P. S. Jan, Collagen–carbon nanotube composite materials as scaffolds in tissue engineering. *J. Biomed. Mater. Res. Part A*, 74A (2005) 489–96.

Index